Differential Geometry

Differential Geometry

Special Issue Editor

Ion Mihai

MDPI • Basel • Beijing • Wuhan • Barcelona • Belgrade

Special Issue Editor
Ion Mihai
University of Bucharest
Romania

Editorial Office
MDPI
St. Alban-Anlage 66
4052 Basel, Switzerland

This is a reprint of articles from the Special Issue published online in the open access journal *Mathematics* (ISSN 2227-7390) from 2017 to 2019 (available at: https://www.mdpi.com/journal/mathematics/special_issues/Differential_Geometry).

For citation purposes, cite each article independently as indicated on the article page online and as indicated below:

LastName, A.A.; LastName, B.B.; LastName, C.C. Article Title. *Journal Name* **Year**, *Article Number*, Page Range.

ISBN 978-3-03921-800-4 (Pbk)
ISBN 978-3-03921-801-1 (PDF)

Contents

About the Special Issue Editor

Ion Mihai obtained his Ph.D. in Mathematics at the Katholieke Universiteit Leuven, Belgium. He is currently Full Professor and Ph.D. supervisor at the University of Bucharest, Romania, Faculty of Mathematics and Computer Science, Department of Mathematics. He has published more than 120 papers in differential geometry and has edited books and conference proceedings in this field. His current research interests include the geometry of complex and contact manifolds, geometry of submanifolds, theory of Chen invariants and Chen inequalities, and statistical manifolds and their submanifolds.

Preface to "Differential Geometry"

Differential geometry is the field of mathematics that is concerned with studies of geometrical structures on differentiable manifolds using techniques of differential calculus, integral calculus, and linear algebra. Starting from some classical examples (e.g., open sets in Euclidean spaces, spheres, tori, projective spaces, and Grassmannians) one may construct new manifolds by using algebraic tools: product of manifolds, quotient spaces, pullback of manifolds by smooth functions, tensor product of submanifolds, and numerous others.

Differential geometry became a field of research in late 19th century, but it is still very relevant due to its applications and the development of new approaches. In order to determine the lengths of curves, areas of surfaces, and volumes of manifolds, the geometers have considered Riemannian manifolds or, more generally, pseudo-Riemannian manifolds. On such manifolds, distinguished vector fields (Killing, conformal, concurrent, torse-forming vector fields) have interesting applications in geometry and relativity.

Curvature invariants are the most natural and most important Riemannian invariants as they play key roles in physics and biology. Among the Riemannian curvature invariants, the most investigated are the sectional curvature, scalar curvature, Ricci curvature, and Chen invariants. Mostly studied are the Riemannian manifolds endowed with certain endomorphisms of their tangent bundles: almost complex, almost product, almost contact, and almost paracontact manifolds. More general manifolds, for instance, affine manifolds and statistical manifolds, are also considered.

On the other hand, the geometry of submanifolds in Riemannian manifolds is an important topic of research in differential geometry. Its origins are in the theory of curves and surfaces in the three-dimensional Euclidean space. Obstructions to the existence of minimal, Lagrangian, slant submanifolds were obtained in terms of their Riemannian curvature invariants.

The purpose of the Special Issue "Differential Geometry" of the journal Mathematics was to provide a collection of papers that reflect modern topics of research and new developments in the field of differential geometry and explore applications in other areas. We are very obliged to the journal Mathematics for the opportunity to publish this book.

Ion Mihai
Special Issue Editor

mathematics

MDPI

Article

The Characterization of Affine Symplectic Curves in $\mathbb{R}^4$

Esra Çiçek Çetin and Mehmet Bektaş *

Department of Mathematics, Faculty of Science, Firat University, 23119 Elazığ, Turkey; esracetincicek@gmail.com

* Correspondence: mbektas@firat.edu.tr; Tel.: +90-424-237-0000

Received: 29 November 2018; Accepted: 18 January 2019; Published: 21 January 2019

Abstract: Symplectic geometry arises as the natural geometry of phase-space in the equations of classical mechanics. In this study, we obtain new characterizations of regular symplectic curves with respect to the Frenet frame in four-dimensional symplectic space. We also give the characterizations of the symplectic circular helices as the third- and fourth-order differential equations involving the symplectic curvatures.

Keywords: symplectic curves; circular helices; symplectic curvatures; Frenet frame

1. Introduction

As the Riemannian geometry involves the length as the fundamental quantity, symplectic geometry involves the directed area, and contact geometry involves the twisting behavior as the fundamental quantities. Since contact geometry is always odd-dimensional and symplectic geometry is always even-dimensional, they are dual in the sense that they have many common results. Hence, studying the twisting behavior in symplectic geometry helps us to obtain connections between these two geometries.

The even-dimensional symplectic geometry has been found in numerous areas of mathematics and physics. It arises as the natural geometry of phase-space in the equations of classical mechanics, which are called Hamilton's equations, and treating mechanical problems in phase-space greatly simplifies the problem [1]. Besides, the symplectic numerical methods are known to be fast and accurate [2–5]. Symplectic geometry also arises in microlocal analysis [6–8], in time series analysis [9,10], analysis of random walks on euclidean graphs [11], and applications of Clifford algebras [12–14].

Geometrical optics has been recognized as a semi-classical limit of wave optics with a small parameter; it has nevertheless been constantly considered as a self-consistent theory for light rays, borrowing much from differential geometry and, more specifically, from Riemannian and symplectic geometries. Geometrical optics provides, indeed, a beautiful link between both previously-mentioned geometries: (i) Light travels along geodesics of an optical medium, a three-dimensional manifold whose Riemannian structure is defined by a refractive index; (ii) The set of all such geodesics is naturally endowed with the structure of four-dimensional symplectic manifolds [15,16].

The aim of this paper is to study some characterization for a special class of symplectic curves called affine symplectic helices, which are a very important tool for both physics and geometric optics. The helix is a symplectic similarity of non-symbolic full toric diversity, whereby algebraic geometry accounts for the effects of uniformity near the focus-focus singularities. The characterization of the helices in different geometries has also been studied by several researchers [17–21]. Proceeding the same way, we study symplectic regular curves, which are parameterized by the symplectic arc length and analyzed by their Frenet-type symplectic frame. In Section 2, we present the preliminaries on the symplectic geometry in terms of isometry groups and inner products. In Section 3, we give the general properties of affine symplectic curves in $\mathbb{R}^4$, which was firstly studied in [22]. Finally, in Section 4, we present the results that we obtain on the characterizations of symplectic curves in $\mathbb{R}^4$ and study symplectic helices.

2. Preliminaries

In the following, we use similar notations and concepts as in [22].

Let $\mathbb{R}^4$ be endowed with standard symplectic form Ω given in global Darboux coordinates: $z = (x^1, x^2, y^1, y^2)$ by

$$\Omega = dx^1 \wedge dy^1 + dx^2 \wedge dy^2. \tag{1}$$

Given two vector fields:

$$\mathbf{u} = x^1 \frac{\partial}{\partial x^1} + x^2 \frac{\partial}{\partial x^2} + y^1 \frac{\partial}{\partial y^1} + y^2 \frac{\partial}{\partial y^2}$$

and:

$$\mathbf{v} = \xi^1 \frac{\partial}{\partial \xi^1} + \xi^2 \frac{\partial}{\partial \xi^2} + \omega^1 \frac{\partial}{\partial \omega^1} + \omega^2 \frac{\partial}{\partial \omega^2}$$

the symplectic form (1) induces a symplectic inner product, which is a non-degenerate, skew-symmetric, bilinear form, on each fiber of tangent bundle $T\mathbb{R}^4$. with:

$$< \mathbf{u}, \mathbf{v} >= \Omega(\mathbf{u}, \mathbf{v}) = \sum_{i=1}^{2} \left(x^i \omega^i - y^i \xi^i \right). \tag{2}$$

The isometry group of the inner product (2) is the 10-dimensional symplectic group $Sp(4) = Sp(4, \mathbb{R}) \subset GL(4, \mathbb{R})$. The Lie algebra $\mathfrak{sp}(4)$ of $Sp(4)$ is the vector space consisting of all 4×4 matrices of the form:

$$\begin{pmatrix} U & V \\ W & -U^T \end{pmatrix}, \tag{3}$$

where U, V, and W are 2×2 matrices satisfying:

$$W = W^T, \ V = V^T.$$

The semi-direct product $G = Sp(4, \mathbb{R}) \ltimes \mathbf{R}^4$ of the symplectic group by the translations is called the group of rigid symplectic motions [22]. Hence, a rigid symplectic motion acting on $z \in \mathbf{R}^4$ with $z \mapsto Az + b$ for $(A, b) \in Sp(4, \mathbb{R})$ is an affine symplectic transformation.

Definition 1. *A symplectic frame is a smooth section of the bundle of linear frames over $\mathbb{R}^4$, which assigns to every point $z \in \mathbb{R}^4$ an ordered basis of tangent vectors a_1, a_2, a_3, a_4 with the property that:*

$$\begin{aligned} \langle a_i, a_j \rangle &= \langle a_{2+i}, a_{2+j} \rangle = 0 \, , \ 1 \le i, j \le 2, \\ \langle a_i, a_{2+j} \rangle &= 0 \, , \ 1 \le i \ne j \le 2, \\ \langle a_i, a_{2+i} \rangle &= 1 \, , \ 1 \le i \le 2. \end{aligned} \tag{4}$$

The structure equations for a symplectic frame are therefore of the form:

$$\begin{aligned} da_i &= \sum_{k=1}^{2} w_{ik} a_k + \sum_{k=1}^{2} \theta_{ik} a_{2+k} \\ da_{2+i} &= \sum_{k=1}^{2} \phi_{ik} a_k - \sum_{k=1}^{2} w_{ki} a_{2+k} \end{aligned} \tag{5}$$

for $1 \leq i \leq 2$. By a consequence of the conditions in (4), the one forms satisfy:

$$\theta_{ij} = \theta ji, \ \ \phi_{ij} = \phi_{ji}. \tag{6}$$

3. General Properties of Affine Symplectic Curves in $\mathbb{R}^4$

We consider parametrized smooth curves $z : I \to \mathbb{R}^4$ defined on an open interval $I \subset \mathbf{R}$. As is customary in classical mechanics, we use the notation $\dot{z}$ to denote differentiation with respect to the parameter t, that is:

$$\dot{z} = \frac{dz}{dt}. \tag{7}$$

Definition 2. *Let $z : I \to \mathbb{R}^4$ be a smooth curve. If the second-order osculating spaces of z satisfy the non-degeneracy condition:*

$$< \dot{z}, \ddot{z} > \neq 0$$

for all $t \in I$, then $z : I \to \mathbb{R}^4$ is called an affine symplectic regular curve.

Definition 3. *Let $t_0 \in I$. The symplectic arc length s of a symplectic regular curve z starting at t_0 is defined by:*

$$s(t) = \int_{t_0}^{t} \langle \dot{z}, \ddot{z} \rangle^{1/3} \, dt \tag{8}$$

for $t \in I$.

We shall note that symplectic arc length may be negative. However, with no loss of generality, we may assume that $< \dot{z}, \ddot{z} >> 0$ throughout the paper.

Taking the exterior differential of the (8), we obtain the symplectic arc length element as:

$$ds = \langle \dot{z}, \ddot{z} \rangle^{1/3} \, dt. \tag{9}$$

In the following, primes are used to denote differentiation with respect to the symplectic arc length derivative operator (9) as:

$$z' = \frac{dz}{ds}.$$

Definition 4. *A symplectic regular curve is parameterized by the symplectic arc length if:*

$$\langle \dot{z}, \ddot{z} \rangle = 1 \tag{10}$$

for all $t \in I$.

Proposition 1. *Every symplectic regular curve can be parameterized by the symplectic arc length.*

Proposition 2. *Let $z : I \to \mathbb{R}^4$ be a symplectic regular curve, which is parameterized by the symplectic arc length, and such that $H_2(s) \neq 0$. Then, the symplectic frame $\{a_1(s), a_2(s), a_3(s), a_4(s)\}$ defined along the image of z satisfies the following structure equations:*

$$\begin{aligned} a_1'(s) &= a_3(s) \\ a_2'(s) &= H_2(s) a_4(s) \\ a_3'(s) &= k_1(s) a_1(s) + a_2(s) \\ a_4'(s) &= a_1(s) + k_2(s) a_2(s), \end{aligned} \tag{11}$$

where $H_2(s), k_1(s), k_2(s)$ are symplectic curvatures of z.

In general, we call the equations in (11) symplectic Frenet equations.

4. The Characterizations of Symplectic Curves in $\mathbb{R}^4$

Definition 5. *Let $z : I \to \mathbb{R}^4$ be a symplectic regular curve, which is parameterized by the symplectic arc length, and $\{a_1(s), a_2(s), a_3(s), a_4(s)\}$ be the Frenet frame of this curve. A symplectic curve z that satisfies the following condition:*

$$\frac{k_1(s)}{k_2(s)} = const.$$

is called a general helix with respect to the Frenet frame.

Example 1. *Let $z : I \to \mathbb{R}^4$ be defined with $z(t) = \left(t, \frac{t^2}{2}, \frac{t^3}{3}, \frac{t^3}{3} + \frac{t^5}{5}\right)$. Since $\Omega(dz, dz) \neq 0$ and:*

$$\frac{k_1(t)}{k_2(t)} = constant$$

with:

$$ds = \langle \dot{z}, \ddot{z} \rangle^{1/3} dt,$$

z is a symplectic polynomial helix.

Example 2. *Let $z : I \to \mathbb{R}^4$ be defined with $z(s) = (\cosh s, 0, \sinh s, 0)$. Since $\Omega(dz, dz) \neq 0$ and:*

$$\frac{k_1(s)}{k_2(s)} = constant$$

with:

$$< \dot{z}, \ddot{z} >= 1,$$

z is a symplectic arc length parameterized circular helix.

Definition 6. *Let $z : I \to \mathbb{R}^4$ be a symplectic regular curve, which is parameterized by the symplectic arc length, and $\{ a_1(s), a_2(s), a_3(s), a_4(s)\}$ be the Frenet frame of z. If both $k_1(s)$ and $k_2(s)$ are positive constants along z, then z is called a circular helix with respect to the Frenet frame.*

Theorem 1. *Let $z(s)$ be a symplectic regular curve, which is parameterized by the symplectic arc length. $z(s)$ is a general helix with respect to the Frenet frame $\{a_1(s), a_2(s), a_3(s), a_4(s)\}$ such that $H_2(s) = const \neq 0$ if and only if:*

$$a_1^{(iv)}(s) = [k_1^{''}(s) + k_1^2(s) + H_2(s)]a_1(s) + 2k_1^{'}(s)a_3(s) \tag{12}$$

Proof. Suppose that $z(s)$ is a general helix with respect to the Frenet frame $\{a_1(s), a_2(s), a_3(s), a_4(s)\}$. Then, from (11), we have:

$$\begin{aligned} a_1^{(iv)}(s) = \; & [k_1^{''}(s) + k_1^2(s) + H_2(s)]a_1(s) + [k_1(s) + \\ & k_2(s)H_2(s)]a_2(s) + 2k_1^{'}(s)a_3(s) + H_2^{'}(s)a_4(s) \end{aligned} \tag{13}$$

Now, $H_2(s) = cons(\neq 0)$, and $z(s)$ is a general helix with respect to the Frenet frame; we suppose that:

$$\frac{k_1(s)}{k_2(s)} = -H_2(s) \tag{14}$$

If we substitute Equation (14) in (13), we obtain (12).

Conversely, let us assume that Equation (12) holds. We show that the curve $z(s)$ is a general helix. From (11), we obtain:

$$a_1(s) = \frac{1}{k_1(s)}[a_3'(s) - a_2(s)] \tag{15}$$

Differentiating covariantly (15), we obtain:

$$a_1'(s) = \left(\frac{-k_1'(s)}{k_1(s)}\right) a_1(s) + \left(\frac{1}{k_1(s)}\right) [a_3''(s) - a_2'(s)] \tag{16}$$

and so:

$$\begin{aligned} a_1''(s) &= \left(\frac{-k_1'(s)}{k_1(s)}\right)' a_1(s) + \left(\frac{-k_1'(s)}{k_1(s)}\right) a_3(s). \\ &+ \left(\frac{1}{k_1(s)}\right)' [a_3''(s) - a_2'(s)] + \left(\frac{1}{k_1(s)}\right) [a_3''' - a_2''] \end{aligned} \tag{17}$$

If we use (7) in (17) and after routine calculations, we have:

$$\frac{H_2'(s)}{k_1(s)} = 0 \tag{18}$$

and:

$$\frac{-H_2(s)k_2(s)}{k_1(s)} = 1. \tag{19}$$

Hence, we obtain $H_2(s) = const.$ and $\frac{k_1(s)}{k_2(s)} = const$. This shows that $z(s)$ is a general helix. □

The hypotheses of Theorem 1 and the definition of a circular helix lead us to the following corollary:

Corollary 1. *Let $z(s)$ be a symplectic regular curve, which is parametrized by the symplectic arc length. $z(s)$ is a circular helix with respect to the Frenet frame $\{a_1(s), a_2(s), a_3(s), a_4(s)\}$ if and only if:*

$$a_1^{(iv)}(s) = \lambda a_1(s), \tag{20}$$

where $\lambda = k_1^2(s) + H_2(s) = const.$

Theorem 2. *Let $z(s)$ be a symplectic regular curve, which is parametrized by the symplectic arc length. $z(s)$ is a circular helix with respect to the Frenet frame $\{a_1(s), a_2(s), a_3(s), a_4(s)\}$ if and only if:*

$$a_4'''(s) = \left(1 + k_2''(s) - k_1(s)k_2(s)\right) a_2(s) + \left(2k_2'(s)H_2(s)\right) a_4(s). \tag{21}$$

Corollary 2. *Let $z(s)$ be a symplectic regular curve, which is parametrized by the symplectic arc length. $z(s)$ is a general helix with respect to the Frenet frame $\{a_1(s), a_2(s), a_3(s), a_4(s)\}$ such that $H_2(s) = const \neq 0$ if and only if:*

$$a_4'''(s) = \mu a_2(s), \tag{22}$$

where $\mu = (1 - k_1(s)k_2(s)) = const.$

Theorem 3. *Let $z(s)$ be a symplectic regular curve, which is parametrized by the symplectic arc length. $z(s)$ is a general helix with respect to the Frenet frame $\{a_1(s), a_2(s), a_3(s), a_4(s)\}$ such that $H_2(s) = const \neq 0$ if and only if:*

$$a_2'''(s) = H_2(s)a_3(s) + H_2(s)K_2'(s)a_2(s) - H_2(s)K_1(s)a_4(s) \tag{23}$$

Corollary 3. *Let $z(s)$ be a symplectic regular curve, which is parametrized by the symplectic arc length. $z(s)$ is a general helix with respect to the Frenet frame $\{a_1(s), a_2(s), a_3(s), a_4(s)\}$ such that $H_2(s) = const \neq 0$ if and only if:*

$$a_2^{'''}(s) = c_1 a_3(s) + c_2 a_4(s) \tag{24}$$

$c_1 = H_2(s) = const.$ *and* $c_2 = H_2(s) K_1(s) = const.$

In the rest of this section, we discuss symplectic regular curves with constant local symplectic invariants. The theorem of Cartan states that the curves with constant symplectic curvatures are precisely the orbits of the one-parameter subgroups of the affine symplectic group in four variables [23,24]. In order to determine such one-parameter subgroups, we shall directly integrate the symplectic Frenet equations of affine symplectic helices. Now, let us consider the symplectic Frenet equations given by (11) with the matrix form as:

$$\frac{d}{ds}\begin{pmatrix} a_1 \\ a_2 \\ a_3 \\ a_4 \end{pmatrix} = \begin{pmatrix} 0 & 0 & 1 & 0 \\ 0 & 0 & 0 & H_2 \\ k_1 & 1 & 0 & 0 \\ 1 & k_2 & 0 & 0 \end{pmatrix} \begin{pmatrix} a_1 \\ a_2 \\ a_3 \\ a_4 \end{pmatrix}, \tag{25}$$

with the constant symplectic curvatures k_1, k_2, H_1. It is well known that the eigenvalues of the Frenet matrix appearing in the right-hand side of (25) are:

$$\mu_1 = \frac{1}{\sqrt{2}}\sqrt{\lambda_1 + \sqrt{\lambda_2}},\ \mu_2 = -\mu_2$$
$$\mu_3 = \frac{1}{\sqrt{2}}\sqrt{\lambda_1 - \sqrt{\lambda_2}},\ \mu_4 = -\mu_3,$$

where $\lambda_1 = k_2 H_2 + k_1$ and $\lambda_2 = (k_2 H_2 - k_1)^2 + 4H_2$ [22].

Now, let us assume that $z : I \to \mathbb{R}^4$ is a symplectic general helix with constant positive curvatures k_1, k_2. Then, by Theorem 1, $k_1 = -k_2 H_2$. Therefore, the eigenvalues of the Frenet matrix appearing in (25) become:

$$\mu_1 = \frac{1}{2}\sqrt[4]{\lambda},\ \mu_2 = -\mu_2$$
$$\mu_3 = \frac{i}{2}\sqrt{\lambda},\ \mu_4 = -\mu_3,$$

where $\lambda = \left(k_1^2 + H_1\right)$ and $i = \sqrt{-1}$. Thus, if $H_1 < -k_1^2$, then the eigenvalues are distinct complex conjugates. Similarly, if $H_1 > -k_1^2$, then the eigenvalues are distinct reals. Depending on the two cases involving symplectic curvatures, we obtain symplectic general helices of the euclidean or hyperbolic type.

5. Conclusions

In our three-dimensional world, the four-dimensional Frenet formulae may seem irrelevant and useless. However, in many areas, including the classical mechanics of physics, the Frenet formulae have been applied. In this study, we study four-dimensional symplectic curves by using the Frenet frames. Our results show that a symplectic helix involves non-zero constant symplectic curvature if and only if the fourth derivative of its first component of the position vector can be described as in Equation (12). Besides, the symplectic circular helices can be characterized directly by the first component of the position vector with the fourth-order derivative.

The characterization of the symplectic helices not only depends on the first component of the position vector. The third derivatives of the second and fourth components of the position vector can be characterized as in Equations (21) and (23). Similarly, symplectic circular helices can be characterized directly by their second and fourth components of the position vector with the third-order derivatives.

Helices are natural twisting structures; hence, studying the symplectic helix may shed light on the connection of contact and symplectic geometries.

Author Contributions: These authors contributed equally to this work.

Funding: This research received no external funding.

Conflicts of Interest: The authors declare no conflicts of interest.

References

1. Libermann, P.; Marle, C.M. *Symplectic Geometry and Analytical Mechanics*; Springer Netherlands: Dordrecht, The Netherlands, 1987; Volume 35.
2. Duistermaat, J.J.; Guillemin, V.W.; Hormander, L.; Vassiliev, D. *Fourier Integral Operators*; Birkhauser: Boston, MA, USA, 1996; Volume 2.
3. Saitoh, I.; Suzuki, Y.; Takahashi, N. The symplectic finite difference time domain method. *IEEE Trans. Magn.* **2001**, *37*, 3251–3254. [CrossRef]
4. Yang, H.W.; Song, H. Symplectic FDTD method study left-handed material electromagnetic characteristics. *Opt.-Int. J. Light Electron Opt.* **2013**, *124*, 1716–1720. [CrossRef]
5. Zhong, S.; Ran, C.; Liu, S. The Optimal Force-Gradient Symplectic Finite-Difference Time-Domain Scheme for Electromagnetic Wave Propagation. *IEEE Trans. Antennas Propag.* **2016**, *64*, 5450–5454. [CrossRef]
6. Akgüller, Ö. Poisson Bracket on Measure Chains and Emerging Poisson Manifold. *Int. J. Appl. Math. Stat.* **2017**, *57*, 56–64.
7. Bach, A. *An Introduction to Semiclassical and Microlocal Analysis*; Springer Science & Business Media: New York, NY, USA, 2013.
8. Grigis, A.; Sjöstrand, J. *Microlocal Analysis for Differential Operators: An Introduction*; Cambridge University Press: Cambridge, UK, 1994; Volume 196.
9. Xie, H.B.; Dokos, S.; Sivakumar, B.; Mengersen, K. Symplectic geometry spectrum regression for prediction of noisy time series. *Phys. Rev. E* **2016**, *93*, 052217. [CrossRef] [PubMed]
10. Xie, H.B.; Guo, T.; Sivakumar, B.; Liew, A.W.C.; Dokos, S. Symplectic geometry spectrum analysis of nonlinear time series. *Proc. R. Soc. A* **2014**, *470*, 20140409. [CrossRef]
11. Cantarella, J.; Shonkwiler, C. The symplectic geometry of closed equilateral random walks in 3-space. *Ann. Appl. Probab.* **2016**, *26*, 549–596. [CrossRef]
12. Binz, E.; De Gosson, M.A.; Hiley, B.J. Clifford Algebras in Symplectic Geometry and Quantum Mechanics. *Found. Phys.* **2013**, *43*, 424–439. [CrossRef]
13. Crumeyrolle, A. *Orthogonal and Symplectic Clifford Algebras: Spinor Structures*; Springer: Dordrecht, The Netherlands, 1990; Volume 57.
14. Da Prato, G.; Zabczyk, J. *Ergodicity for Infinite Dimensional Systems*; Cambridge University Press: Cambridge, UK, 1996; Volume 229.
15. Duval, C.; Horvath, Z.; Horváthy, P.A. Geometrical spinoptics and the optical Hall effect. *J. Geom. Phys.* **2007**, *57*, 925–941. [CrossRef]
16. Guillemin, V.; Sternberg, S. *Symplectic Techniques in Physics*; Cambridge University Press: Cambridge, UK, 1990.
17. Bektaş, M. On a characterization of null helix. *Bull. Inst. Math. Acad. Sin.* **2001**, *29*, 71–78.
18. Bektaş, M. On characterizations of general helices for ruled surfaces in the pseudo-Galilean space G_3^1-(Part-I). *J. Math. Kyoto Univ.* **2004**, *44*, 523–528. [CrossRef]
19. Ekmekçi, N.; Hacısalihoğlu, H. On Helicec of a Lorentzian Manifold. *Commun. Math. Stat.* **1995**, *42*, 43–55.
20. Yılmaz, M.Y.; Bektaş, M. Helices of the 3-dimensional Finsler manifold. *J. Adv. Math. Stud.* **2009**, *2*, 107–113.
21. Valiquette, F. Geometric affine symplectic curve flows in $\mathbb{R}^4$. *Differ. Geom. Its Appl.* **2012**, *30*, 631–641. [CrossRef]
22. Kamran, N.; Olver, P.; Tenenblat, K. Local symplectic invariants for curves. *Commun. Contemp. Math.* **2009**, *11*, 165–183. [CrossRef]

23. Fels, M.; Olver, P.J. Moving coframes: I. A practical algorithm. *Acta Appl. Math.* **1998**, *51*, 161–213. [CrossRef]
24. Fels, M.; Olver, P.J. Moving coframes: II. Regularization and theoretical foundations. *Acta Appl. Math.* **1999**, *55*, 127–208. [CrossRef]

mathematics

MDPI

Article

Euclidean Submanifolds via Tangential Components of Their Position Vector Fields

Bang-Yen Chen

Department of Mathematics, Michigan State University, East Lansing, MI 48824-1027, USA; bychen@math.msu.edu; Tel.: +1-517-515-9087

Academic Editor: Ion Mihai
Received: 6 September 2017; Accepted: 10 October 2017; Published: 16 October 2017

Abstract: The position vector field is the most elementary and natural geometric object on a Euclidean submanifold. The position vector field plays important roles in physics, in particular in mechanics. For instance, in any equation of motion, the position vector $\mathbf{x}(t)$ is usually the most sought-after quantity because the position vector field defines the motion of a particle (i.e., a point mass): its location relative to a given coordinate system at some time variable t. This article is a survey article. The purpose of this article is to survey recent results of Euclidean submanifolds associated with the tangential components of their position vector fields. In the last section, we present some interactions between torqued vector fields and Ricci solitons.

Keywords: Euclidean submanifold; position vector field; concurrent vector field; concircular vector field; rectifying submanifold; T-submanifolds; constant ratio submanifolds; Ricci soliton

1. Introduction

For an n-dimensional submanifold M in the Euclidean m-space $\mathbb{E}^m$, the most elementary and natural geometric object is the position vector field $\mathbf{x}$ of M. The position vector is a Euclidean vector $\mathbf{x} = \overrightarrow{OP}$ that represents the position of a point $P \in M$ in relation to an arbitrary reference origin $O \in \mathbb{E}^m$.

The position vector field plays important roles in physics, in particular in mechanics. For instance, in any equation of motion, the position vector $\mathbf{x}(t)$ is usually the most sought-after quantity because the position vector field defines the motion of a particle (i.e., a point mass): its location relative to a given coordinate system at some time variable t. The first and the second derivatives of the position vector field with respect to time t give the velocity and acceleration of the particle.

For a Euclidean submanifold M of a Euclidean m-space, there is a natural decomposition of the position vector field $\mathbf{x}$ given by:

$$\mathbf{x} = \mathbf{x}^T + \mathbf{x}^N, \tag{1}$$

where $\mathbf{x}^T$ and $\mathbf{x}^N$ are the tangential and the normal components of $\mathbf{x}$, respectively. We denote by $|\mathbf{x}^T|$ and $|\mathbf{x}^N|$ the lengths of $\mathbf{x}^T$ and of $\mathbf{x}^N$, respectively. Clearly, we have $|\mathbf{x}^N| = \sqrt{|\mathbf{x}|^2 - |\mathbf{x}^T|^2}$. In [1], the author provided a survey on several topics in differential geometry associated with position vector fields on Euclidean submanifolds.

In this paper, we discuss Euclidean submanifolds M whose tangential components $\mathbf{x}^T$ admit some special properties such as concurrent, concircular, torse-forming, etc. Moreover, we will also discuss constant-ratio submanifolds, as well as Ricci solitons on Euclidean submanifolds with the potential fields of the Ricci solitons coming from the tangential components of the position vector fields. In the last section, we present some interactions between torqued vector fields and Ricci solitons.

2. Preliminaries

Let $x : M \to \mathbb{E}^m$ be an isometric immersion of a Riemannian manifold M into a Euclidean m-space $\mathbb{E}^m$. For each point $p \in M$, we denote by T_pM and $T_p^{\perp}M$ the tangent space and the normal space of M at p, respectively.

Let ∇ and $\tilde{\nabla}$ denote the Levi–Civita connections of M and $\mathbb{E}^m$, respectively. Then, the formulas of Gauss and Weingarten are given respectively by (cf. [2–6]):

$$\tilde{\nabla}_X Y = \nabla_X Y + h(X,Y), \tag{2}$$

$$\tilde{\nabla}_X \xi = -A_\xi X + D_X \xi, \tag{3}$$

for vector fields X, Y tangent to M and ξ normal to M, where h is the second fundamental form, D the normal connection and A the shape operator of M.

At a given point $p \in M$, the first normal space of M in $\mathbb{E}^m$, denoted by $\operatorname{Im} h_p$, is the subspace given by:

$$\operatorname{Im} h_p = \operatorname{Span}\{h(X,Y) : X, Y \in T_pM\}. \tag{4}$$

For each normal vector ξ at p, the shape operator A_ξ is a self-adjoint endomorphism of T_pM. The second fundamental form h and the shape operator A are related by:

$$\langle A_\xi X, Y\rangle = \langle h(X,Y), \xi\rangle, \tag{5}$$

where $\langle\ ,\ \rangle$ is the inner product on M, as well as on the ambient Euclidean space. The covariant derivative $\bar{\nabla} h$ of h with respect to the connection on $TM \oplus T^{\perp}M$ is defined by:

$$(\bar{\nabla}_X h)(Y,Z) = D_X(h(Y,Z)) - h(\nabla_X Y, Z) - h(Y, \nabla_X Z). \tag{6}$$

For a given point $p \in M$, we put:

$$\operatorname{Im}(\bar{\nabla} h_p) = \{\bar{\nabla}_X h)(Y,Z) : X, Y, Z \in T_pM\}. \tag{7}$$

The subspace $\operatorname{Im} \bar{\nabla} h_p$ is called the second normal space at p.

The equation of Gauss of M in $\mathbb{E}^m$ is given by:

$$R(X,Y;Z,W) = \langle h(X,W), h(Y,Z)\rangle - \langle h(X,Z), h(Y,W)\rangle \tag{8}$$

for X, Y, Z, W tangent to M, where R is the Riemann curvature tensors of M defined by:

$$R(X,Y;Z,W) = \langle \nabla_X \nabla_Y Z, W\rangle - \langle \nabla_Y \nabla_X Z, W\rangle - \left\langle \nabla_{[X,Y]} Z, W \right\rangle.$$

The equation of Codazzi is:

$$(\bar{\nabla}_X h)(Y,Z) = (\bar{\nabla}_Y h)(X,Z). \tag{9}$$

The mean curvature vector H of a submanifold M is defined by:

$$H = \frac{1}{n}\operatorname{trace} h, \quad n = \dim M. \tag{10}$$

A Riemannian manifold is called a flat space if its curvature tensor R vanishes identically. Further, a submanifold M is called totally umbilical (respectively, totally geodesic) if its second fundamental form h satisfies $h(X,Y) = \langle X,Y\rangle H$ identically (respectively, $h = 0$ identically).

A hypersurface of a Euclidean $(n+1)$-space $\mathbb{E}^{n+1}$ is called a quasi-umbilical hypersurface if its shape operator has an eigenvalue κ of multiplicity $mult(\kappa) \geq n-1$ (cf. ([2], p. 147)). On the subset U of M on which $mult(\kappa) = n-1$, an eigenvector with eigenvalue of multiplicity one is called a distinguished direction of the quasi-umbilical hypersurface.

The following lemmas can be found in [7].

Lemma 1. *Let $x : M \to \mathbb{E}^m$ be an isometric immersion of a Riemannian n-manifold into a Euclidean m-space $\mathbb{E}^m$. Then, $\mathbf{x} = \mathbf{x}^T$ holds identically if and only if M is a conic submanifold with the vertex at the origin.*

Lemma 2. *Let $x : M \to \mathbb{E}^m$ be an isometric immersion of a Riemannian n-manifold into $\mathbb{E}^m$. Then, $\mathbf{x} = \mathbf{x}^N$ holds identically if and only if M lies in a hypersphere centered at the origin.*

In view of Lemmas 1 and 2, we make the following.

Definition 1. *A submanifold M of $\mathbb{E}^m$ is called proper if it satisfies $\mathbf{x} \neq \mathbf{x}^T$ and $\mathbf{x} \neq \mathbf{x}^N$ almost everywhere.*

3. Euclidean Submanifolds with Constant $|\mathbf{x}^T|$ or Constant $|\mathbf{x}^N|$

Euclidean submanifolds with constant $|\mathbf{x}^T|$ are called T-constant submanifolds in [8]. These submanifolds were first introduced and studied by the author in [8].

One important property of a T-constant proper hypersurface M is that the tangential component $\mathbf{x}^T$ of the position vector field $\mathbf{x}$ of M defines a principal direction for the hypersurface. Moreover, the normal component $\mathbf{x}^N$ of M is nowhere zero (see, ([8], p. 66)).

T-constant Euclidean proper submanifolds were classified in [8] as follows.

Theorem 1. *Let $x : M \to \mathbb{E}^m$ be an isometric immersion of a Riemannian n-manifold into the Euclidean m-space. Then, M is a T-constant proper submanifold if and only if there exist real numbers a, b and local coordinate systems $\{s, u_2, \ldots, u_n\}$ on M such that the immersion x is given by:*

$$x(s, u_2, \ldots, u_n) = \sqrt{a^2 + b + 2as}\, Y(s, u_2, \ldots, u_n), \tag{11}$$

where $Y = Y(s, u_2, \ldots, u_n)$ satisfies the following conditions:

(a) *$Y = Y(s, u_2, \ldots, u_n)$ lies in the unit hypersphere $S^{m-1}(1)$,*
(b) *the coordinate vector field Y_s is perpendicular to coordinate vector fields $Y_{u_2}, \ldots, Y_{u_n}$ and*
(c) *Y_s satisfies $|Y_s| = \sqrt{b + 2as}/(a^2 + b + 2as)$.*

Now, we provide some examples of T-constant proper hypersurfaces in $\mathbb{R}^{n+1}$.

Example 1. *For a given real number $a > 0$ and for $s > 0$, we define $Y = Y(s, u_2, \ldots, u_n)$ by:*

$$\begin{aligned} Y = \frac{1}{\sqrt{a^2 + 2as}} \Bigg(& a \sin\left(\frac{\sqrt{2as}}{a}\right) - \sqrt{2as} \cos\left(\frac{\sqrt{2as}}{a}\right), \\ & \left\{ a \cos\left(\frac{\sqrt{2as}}{a}\right) + \sqrt{2as} \sin\left(\frac{\sqrt{2as}}{a}\right) \right\} \prod_{j=2}^{n} \cos u_j, \\ & \left\{ a \cos\left(\frac{\sqrt{2as}}{a}\right) + \sqrt{2as} \sin\left(\frac{\sqrt{2as}}{a}\right) \right\} \sin u_2, \ldots, \\ & \left\{ a \cos\left(\frac{\sqrt{2as}}{a}\right) + \sqrt{2as} \sin\left(\frac{\sqrt{2as}}{a}\right) \right\} \sin u_n \prod_{j=2}^{n-1} \cos u_j \Bigg) \end{aligned} \tag{12}$$

in $\mathbb{E}^{n+1}$. Then, $|Y| = 1$, and $Y = Y(s, u_2, \dots, u_n)$ satisfies the conditions (a), (b) and (c) of Theorem 1. An easy computation shows that:

$$x = \sqrt{a^2 + 2as}\, Y(s, u_2, \dots, u_n) \tag{13}$$

satisfies $|\mathbf{x}^T| = a$. Thus, (13) defines a proper T-constant submanifold in $\mathbb{E}^{n+1}$.

Similarly, one may also consider Euclidean submanifolds with constant $|\mathbf{x}^N|$. Such submanifolds are called N-constant submanifolds in [8].

Proper N-constant Euclidean submanifolds were classified in [8] as follows.

Theorem 2. *Let $x : M \to \mathbb{E}^m$ be an isometric immersion of a Riemannian n-manifold into the Euclidean m-space. Then, M is an N-constant proper submanifold if and only if there exist a positive number c and local coordinate systems $\{s, u_2, \dots, u_n\}$ on M such that the immersion x is given by:*

$$x(s, u_2, \dots, u_n) = \sqrt{s^2 + c^2}\, Y(s, u_2, \dots, u_n), \tag{14}$$

where $Y = Y(s, u_2, \dots, u_n)$ satisfies the conditions:

(1) *$Y = Y(s, u_2, \dots, u_n)$ lies in the unit hypersphere $S^{m-1}(1)$,*
(2) *Y_s is perpendicular to coordinate vector fields $Y_{u_2}, \dots, Y_{u_n}$ and*
(3) *Y_s satisfies $|Y_s| = c/(s^2 + c^2)$.*

Here are some examples of N-constant proper hypersurfaces of $\mathbb{E}^{n+1}$.

Example 2. *For a given positive numbers c, we define:*

$$Y = \frac{1}{\sqrt{s^2 + c^2}} \left(c, s \prod_{j=2}^{n} \cos u_j, s \sin u_2, \dots, s \sin u_n \prod_{j=2}^{n-1} \cos u_j \right) \tag{15}$$

in $\mathbb{E}^{n+1}$. Then, $\langle Y, Y \rangle = 1$, and $Y = Y(s, u_2, \dots, u_n)$ satisfies the conditions (1), (2) and (3) of Theorem 2.
An easy computation shows that:

$$x = \sqrt{s^2 + c^2}\, Y(s, u_2, \dots, u_n) \tag{16}$$

satisfies $\langle x^N, x^N \rangle = c^2$, which provides an example of a proper N-constant submanifold.

4. Euclidean Submanifolds with Constant Ratio $|\mathbf{x}^T| : |\mathbf{x}^N|$

Euclidean submanifolds with the ratio $|\mathbf{x}^T| : |\mathbf{x}^N|$ being constant are called constant ratio submanifolds. The study of such submanifolds was initiated by the author in [9,10].

As we mentioned in [1], constant-ratio curves in a plane are exactly the equiangular curves in the sense of D'Arcy Thompson's biology theory on growth and form [11]. Thus, constant-ratio submanifolds can be regarded as a higher dimensional version of Thompson's equiangular curves. For this reason, constant-ratio submanifolds are also known in some literature as equiangular submanifolds (see, e.g., [12,13]).

Constant-ratio submanifolds were completely classified by the author in [9,10] as follows.

Theorem 3. *Let $x : M \to \mathbb{E}^m$ be an isometric immersion of a Riemannian n-manifold into the Euclidean m-space. Then, M is a constant-ratio proper submanifold if and only if there exists a number $b \in (0, 1)$ and local coordinate systems $\{s, u_2, \dots, u_n\}$ on M such that the immersion x is given by:*

$$x(s, u_2, \dots, u_n) = bs\, Y(s, u_2, \dots, u_n), \tag{17}$$

where $Y = Y(s, u_2, \dots, u_n)$ satisfies the conditions:

(a) $Y = Y(s, u_2, \ldots, u_n)$ *lies in the unit hypersphere* $S^{m-1}(1)$,
(b) Y_s *is perpendicular to* $Y_{u_2}, \ldots, Y_{u_n}$ *and*
(c) $|Y_s| = \sqrt{1-b^2}/(bs)$.

We give the following examples of constant-ration hypersurfaces.

Example 3. *Let b be a real number in $(0,1)$ and $s > 0$. We define:*

$$Y(s, u_2, \ldots, u_n) = \left(\sin\left(\frac{\sqrt{1-b^2}}{b}\ln s\right), \cos\left(\frac{\sqrt{1-b^2}}{b}\ln s\right)\prod_{j=2}^{n}\cos u_j,\right.$$
$$\left.\cos\left(\frac{\sqrt{1-b^2}}{b}\ln s\right)\sin u_2, \ldots, \cos\left(\frac{\sqrt{1-b^2}}{b}\ln s\right)\sin u_n\prod_{j=2}^{n-1}\cos u_j\right)$$

in $\mathbb{E}^{n+1}$. Then, $|Y| = 1$, and $Y = Y(s, u_2, \ldots, u_n)$ is a local parametrization of the unit sphere S^n. Moreover, $Y(s, u_2, \ldots, u_n)$ satisfies Conditions (b) and (c) of Theorem 3.

An easy computation shows that $x(s, u_2, \ldots, u_n) = bsY(s, u_2, \ldots, u_n)$ satisfies $|x| = bs$ and $|x^T| = b^2 s$. Hence, $|x^T| = b|x|$. Consequently, x defines a constant-ratio hypersurface in $\mathbb{E}^{n+1}$.

Remark 1. *Constant-ratio curves also relate to the motion in a central force field that obeys the inverse-cube law. In fact, the trajectory of a mass particle subject to a central force of attraction located at the origin that obeys the inverse-cube law is a curve of constant-ratio. The inverse-cube law was originated from Sir Isaac Newton (1642–1727) in his letter sent on 13 December 1679 to Robert Hooke (1635–1703). This letter is of great historical importance since it reveals the state of Newton's development of dynamics at that time (see, for instance, [14,15], pp. 266–271, [16,17], Book I, Section II, Proposition IX).*

Let ρ denote the distance function of a submanifold M in $\mathbb{E}^m$, i.e., $\rho = |\mathbf{x}|$. It was proven in [18] that the Euclidean submanifold M is of constant-ratio if and only if the gradient of the distance function ρ has constant length.

Remark 2. *Constant ratio submanifolds are related to the notion of convolution manifolds introduced by the author in [18,19], as well.*

5. Rectifying Euclidean Submanifolds with Concurrent $\mathbf{x}^T$

Let $\alpha : I \to \mathbb{E}^3$ be a unit speed curve in the Euclidean three-space $\mathbb{E}^3$ with Frenet-Serret apparatus $\{\kappa, \tau, T, N, B\}$, where κ, τ, T, N and B denote the curvature, the torsion, the unit tangent T, the unit principal normal N and the unit binormal of α, respectively. Then, α is called a Frenet curve if the curvature and torsion of α satisfy $\kappa > 0$ and $\tau \neq 0$.

The famous Frenet formulas of α are given by:

$$\begin{cases} \mathbf{t}' = \kappa\mathbf{n}, \\ \mathbf{n}' = -\kappa\mathbf{t} + \tau\mathbf{b}, \\ \mathbf{b}' = -\tau\mathbf{n}. \end{cases} \tag{18}$$

At each point of the curve, the planes spanned by $\{\mathbf{t}, \mathbf{n}\}$, $\{\mathbf{t}, \mathbf{b}\}$ and $\{\mathbf{n}, \mathbf{b}\}$ are known as the osculating plane, the rectifying plane and the normal plane, respectively.

It is well known in elementary differential geometry that a curve in $\mathbb{E}^3$ lies in a plane if its position vector $\mathbf{x}$ lies in its osculating plane at each point; and it lies on a sphere if its position vector lies in its normal plane at each point. In view of these basic facts, the author defined a rectifying curve in $\mathbb{E}^3$ as a Frenet curve whose position vector field always lie in its rectifying plane [20]. Moreover, he completely classified in [20] rectifying curves in $\mathbb{E}^3$. Furthermore, he proved in [21] that a curve on a general cone (not necessarily a circular one) in $\mathbb{E}^3$ is a geodesic if and only if it is a rectifying curve or an open portion of a ruling of the cone. In [22], several interesting links between rectifying curves, centrodes and extremal curves were established by B.-Y. Chen and F. Dillen. Some further results in this respect were also obtained recently in [23,24].

Clearly, it follows from the definition of a rectifying curve $\alpha : I \to \mathbb{E}^3$ that the position vector field $\mathbf{x}$ of α satisfies:

$$\mathbf{x}(s) = \lambda(s)\mathbf{t}(s) + \mu(s)\mathbf{b}(s) \tag{19}$$

for some functions λ and μ.

For a Frenet curve $\gamma : I \to \mathbb{E}^3$, the first normal space of γ at s_0 is the line spanned by the principal normal vector $\mathbf{n}(s_0)$. Hence, the rectifying plane of γ at s_0 is nothing but the plane orthogonal to the first normal space at s_0. For this reason, for a submanifold M of $\mathbb{E}^m$ and a point $p \in M$, we call the subspace of $T_p\mathbb{E}^m$ the rectifying space of M at p if it is the orthogonal complement to the first normal space $\operatorname{Im}\sigma_p$.

According to [7], a submanifold M of a Euclidean m-space $\mathbb{E}^m$ is called a rectifying submanifold if the position vector field $\mathbf{x}$ of M always lies in its rectifying space. In other words, M is called a rectifying submanifold if and only if:

$$\langle \mathbf{x}(p), \operatorname{Im} h_p \rangle = 0 \tag{20}$$

holds for each point $p \in M$. A non-trivial vector field Z on a Riemannian manifold M is called concurrent if it satisfies $\nabla_X Z = X$ for any vector X tangent to M, where ∇ is the Levi–Civita connection of M (cf. [25–28]).

The following results on rectifying submanifolds were proven in [7,29].

Theorem 4. *If M is a proper submanifold of $\mathbb{E}^m$, then M is a rectifying submanifold if and only if $\mathbf{x}^T$ is a concurrent vector field on M.*

Theorem 5. *A proper hypersurface M of $\mathbb{E}^{n+1}$ is rectifying if and only if M is an open portion of a hyperplane L of $\mathbb{E}^{n+1}$ with $o \notin L$, where zero denotes the origin of $\mathbb{E}^{n+1}$.*

Theorem 6. *Let M be a rectifying proper submanifold of $\mathbb{E}^m$. If $m \geq 2 + \dim M$, then with respect to some suitable local coordinate systems $\{s, u_2, \ldots, u_n\}$ on M, the immersion x of M in $\mathbb{E}^m$ takes the form:*

$$x(s, u_2, \ldots, u_n) = \sqrt{s^2 + c^2}\, Y(s, u_2, \ldots, u_n), \;\; \langle Y, Y \rangle = 1, \; c > 0, \tag{21}$$

such that the metric tensor g_Y of the spherical submanifold defined by Y satisfies:

$$g_Y = \frac{c^2}{(s^2 + c^2)^2} ds^2 + \frac{s^2}{s^2 + c^2} \sum_{i,j=2}^{n} g_{ij}(u_2, \ldots, u_n) du_i du_j. \tag{22}$$

Conversely, the immersion defined by (21) and (22) is a rectifying proper submanifold.

Remark 3. *For the pseudo-Euclidean version of Theorem 5, see [30].*

6. Euclidean Submanifolds with Concircular $\mathbf{x}^T$

A non-trivial vector field Z on a Riemannian manifold M is called a concircular vector field if it satisfies (cf., e.g., [5,31,32]):

$$\nabla_X Z = \varphi X, \;\; X \in TM, \tag{23}$$

where φ is a smooth function on M, called the concircular function. Obviously, a concircular vector field with $\varphi = 1$ is a concurrent vector field. For simplicity, we call a Euclidean submanifold with concircular $\mathbf{x}^T$ a circular submanifold.

The following result from [33] classifies concircular submanifolds completely.

Theorem 7. *Let M be a proper submanifold of a Euclidean m-space $\mathbb{E}^m$ with origin zero. If $n = \dim M \geq 2$, then M is a concircular submanifold if and only if one of the following three cases occurs:*

(a) *M is an open portion of a linear n-subspace L^n of $\mathbb{E}^m$ such that $o \notin L$.*
(b) *M is an open portion of a hypersphere S^n of a linear $(n+1)$-subspace L^{n+1} of $\mathbb{E}^m$ such that the origin of $\mathbb{E}^m$ is not the center of S^n.*
(c) *$m \geq n+2$. Moreover, with respect to some suitable local coordinate systems $\{s, u_2, \ldots, u_n\}$ on M, the immersion x of M in $\mathbb{E}^m$ takes the following form:*

$$x(s, u_2, \ldots, u_n) = \sqrt{2\rho}\, Y(s, u_2, \ldots, u_n), \quad \langle Y, Y \rangle = 1, \tag{24}$$

where $Y : M \to S_o^{m-1}(1) \subset \mathbb{E}^m$ is an immersion of M into the unit hypersphere $S_o^{m-1}(1)$ such that the induced metric g_Y via Y is given by:

$$g_Y = \frac{2\rho - \rho'^2}{4\rho^2} ds^2 + \frac{\rho'^2}{2\rho} \sum_{i,j=2}^{n} g_{ij}(u_2, \ldots, u_n) du_i du_j. \tag{25}$$

where $\rho = \rho(s)$ satisfies $2\rho > \rho'^2 > 0$ on an open interval I.

Next, we provide one explicit example of a concircular surface in $\mathbb{E}^4$.

Example 4. *If we choose $n = 2$ and $\rho(s) = \frac{3}{8}s^2$, then (33) reduces to:*

$$g_Y = \frac{1}{3s^2} ds^2 + \frac{3}{4} du^2. \tag{26}$$

Let us define $Y : I_1 \times I_2 \to S_o^3(1) \subset \mathbb{E}^4$ to be the map of $I_1 \times I_2$ into $S_o^3(1)$ given by:

$$Y(s, u) = \frac{1}{\sqrt{2}} \left(\cos\left(\tfrac{\sqrt{2}}{\sqrt{3}} \ln s\right), \sin\left(\tfrac{\sqrt{2}}{\sqrt{3}} \ln s\right), \cos\left(\tfrac{\sqrt{6}}{2} u\right), \sin\left(\tfrac{\sqrt{6}}{2} u\right) \right). \tag{27}$$

Then, the induced metric tensor of $I_1 \times I_2$ via the map Y is given by (34). Therefore, $P^2 = (I_1 \times I_2, g_Y)$ with the induced metric tensor g_Y being a flat surface.

Now, consider $x(s, u) : I_1 \times I_2 \to \mathbb{E}^4$ given by $x(s, u) = F(s) Y(s, u)$, i.e.,

$$x(s, u) = \frac{\sqrt{3}s}{2\sqrt{2}} \left(\cos\left(\tfrac{\sqrt{2}}{\sqrt{3}} \ln s\right), \sin\left(\tfrac{\sqrt{2}}{\sqrt{3}} \ln s\right), \cos\left(\tfrac{\sqrt{6}}{2} u\right), \sin\left(\tfrac{\sqrt{6}}{2} u\right) \right). \tag{28}$$

Then, it is easy to verify that the induced metric via x is:

$$g = ds^2 + \frac{9}{16} s^2 du^2. \tag{29}$$

Hence, the Levi–Civita connection of $M = (I_1 \times I_2, g)$ satisfies:

$$\nabla_{\frac{\partial}{\partial s}} \frac{\partial}{\partial s} = 0, \quad \nabla_{\frac{\partial}{\partial u}} \frac{\partial}{\partial s} = \frac{1}{s} \frac{\partial}{\partial u}. \tag{30}$$

Using (28) and (29), it is easy to verify that the tangential component $\mathbf{x}^T = \frac{3}{4}s\frac{\partial}{\partial s}$ of the position vector field $\mathbf{x}$ is a concircular vector field satisfying $\nabla_Z \mathbf{x}^T = \frac{3}{4}Z$ for $Z \in TM$. Consequently, M is a concircular surface in $\mathbb{E}^4$.

Remark 4. *Concircular vector fields play some important roles in general relativity. For instance, it was proven in [34] that a Lorentzian manifold is a generalized Robertson–Walker spacetime if and only if it admits a timelike concircular vector field. For the most recent surveys on generalized Robertson–Walker spacetimes, see [5,35].*

Remark 5. *It was proven in [36] that every Kaehler manifold M (or more generally, pseudo-Kaehler manifold) with* $\dim_{\mathbb{C}} M > 1$ *does not admit a non-trivial concircular vector field.*

7. Euclidean Submanifolds with Torse-Forming $\mathbf{x}^T$

In [37], K. Yano extended concurrent and concircular vector fields to torse-forming vector fields. According to K. Yano, a vector field v on a Riemannian (or pseudo-Riemannian) manifold M is called a torse-forming vector field if it satisfies:

$$\nabla_X v = \varphi X + \alpha(X)v, \quad \forall X \in TM, \tag{31}$$

for a function φ and a one-form α on M. The one-form α is called the generating form, and the function φ is called the conformal scalar (see [38]). A torqued vector field is a torse-forming vector field v satisfying (31) with $\alpha(v) = 0$ (see [39,40]).

Generalized Robertson–Walker (GRW) spacetimes were introduced by L. J. Alías, A. Romero and M. Sánchez in [41]. The first author proved in [34] that a Lorentzian manifold is a GRW spacetime if and only if it admits a time-like concircular vector field. For further results in this respect, see an excellent survey on GRW spacetimes by C. A. Mantica and L. G. Molinari [35] (see also [5]).

Twisted products are natural extensions of warped products in which the warping functions were replaced by twisting functions (cf. [5,42]). It was proven in [39] that a Lorentzian manifold is a twisted space of the form $I \times_\lambda F$ with time-like base I if and only if it admits a time-like torqued vector field. Recently, C. A. Mantica and L. G. Molinari proved in [43] that such a Lorentzian twisted space can also be characterized as a Lorentzian manifold admitting a torse-forming time-like unit vector field.

Before we state the results for Euclidean hypersurfaces with torse-forming x^T, we give the following simple link between Hessian of functions and torse-forming vector fields.

Theorem 8. *Let f be a non-constant function on a Riemannian manifold M. Then, the gradient ∇f of f is a torse-forming vector field if and only if the Hessian H^f satisfies:*

$$H^f = \varphi g + \gamma df \otimes df \tag{32}$$

where φ and γ are functions on M.

Proof. Let f be a non-constant function on a Riemannian manifold M. Assume that the gradient ∇f of f is a torse-forming vector field so that:

$$\nabla_X(\nabla f) = \varphi X + \alpha(X)\nabla f \tag{33}$$

for some function φ and one-form α on M. Then, for any vector fields X, Y on M, the Hessian H^f of f satisfies:

$$\begin{aligned} H^f(X,Y) &= XYf - (\nabla_X Y)f = X\langle Y, \nabla f\rangle - \langle \nabla_X Y, \nabla f\rangle \\ &= \langle Y, \nabla_X(\nabla f)\rangle = \varphi\langle Y, X\rangle + \alpha(X)\langle Y, \nabla f\rangle \\ &= \varphi\langle Y, X\rangle + \alpha(X)df(Y). \end{aligned} \tag{34}$$

Since the Hessian $H^f(X,Y)$ is symmetric in X and Y, we derive from (34) that:

$$\begin{aligned}\langle Y, \nabla_X(\nabla f)\rangle &= X\langle Y, \nabla f\rangle - \langle \nabla_X Y, \nabla f\rangle = H^f(X,Y) \\ &= \langle Y, \varphi, X + \gamma df(X)\nabla f\rangle\end{aligned}$$

for vector field X, Y. Therefore, we obtain (32) with $\alpha = \gamma df$. Consequently, the gradient ∇f of f is a torse-forming vector field. □

The following corollary is an easy consequence of Theorem 8.

Corollary 1. *Let f be a non-constant function on a Riemannian manifold M. If the gradient ∇f of f is a torqued vector field, then it is a concircular vector field on M.*

Remark 6. *Theorem 8 extends Lemma 4.1 of [31].*

Next, we present the following results from [44] for Euclidean hypersurfaces with torse-forming $\mathbf{x}^T$.

Proposition 1. *Let M be a proper hypersurface of $\mathbb{E}^m$. If the tangential component $\mathbf{x}^T$ of the position vector field $\mathbf{x}$ of M is a torse-forming vector field, then M is a quasi-umbilical hypersurface with $\mathbf{x}^T$ as its distinguished direction.*

For quasi-umbilical hypersurfaces in $\mathbb{E}^m$ we refer to [2,45].

A rotational hypersurface $M = \gamma \times S^{n-1}$ in $\mathbb{E}^{n+1}$ is an $O(n-1)$-invariant hyper-surface, where S^{n-1} is a Euclidean sphere and:

$$\gamma(x) = (x, g(x)),\ \ g(x) > 0,\ \ x \in I, \tag{35}$$

is a plane curve (the profile curve) defined on an open interval I and the x-axis is called the axis of rotation. The rotational hypersurface M can expressed as:

$$\mathbf{x} = (u, g(u)y_1, \cdots, g(u)y_n)\ \text{ with }\ y_1^2 + \cdots + y_n^2 = 1. \tag{36}$$

The hypersurfaces is called a spherical cylinder if its profile curve γ is a horizontal line segment (i.e., $g = constant \neq 0$). Additionally, it is called a spherical cone if γ is a non-horizontal line segment (i.e., $g = cu$, $0 \neq c \in \mathbf{R}$). For simplicity, we only consider rotational hypersurfaces M, which contain no open parts of hyperspheres, spherical cylinders or spherical cones.

A torse-forming vector field v is called proper torse-forming if the one-form α in (31) is nowhere zero on a dense open subset of M.

The simple link between rotational hypersurfaces and torse-forming $\mathbf{x}^T$ is the following.

Theorem 9. *Let M be a proper hypersurface of $\mathbb{E}^{n+1}$ with $n \geq 3$. Then, the tangential component $\mathbf{x}^T$ of the position vector field $\mathbf{x}$ of M is a proper torse-forming vector field if and only if M is an open part of a rotational hypersurface whose axis of rotation contains the origin* [44].

8. Rectifying Submanifolds of Riemannian Manifolds

In [39], the notion of rectifying submanifolds of Euclidean spaces was extended to rectifying submanifolds of Riemannian manifolds.

Definition 2. *Let V be a non-vanishing vector field on a Riemannian manifold $\tilde{M}$, and let M be a submanifold of $\tilde{M}$ such that the normal component V^N of V is nowhere zero on M. Then, M is called a rectifying submanifold (with respect to V) if and only if:*

$$\langle V(p), \operatorname{Im} h_p\rangle = 0 \tag{37}$$

holds at each $p \in M$.

Definition 3. *A submanifold M of a Riemannian manifold* $\tilde{M}$ *is said to be twisted if:*

$$\operatorname{Im} \bar{\nabla} h_p \not\subseteq \operatorname{Im} h_p \tag{38}$$

holds at each point $p \in M$.

A vector field on a Riemannian manifold M is called a gradient vector field if it is the gradient ∇f of some function f on M.

In terms of gradient vector fields, Corollary 1 can be restated as the follows.

Proposition 2. *If a torqued vector field on a Riemannian manifold M is a gradient vector field, then it is a concircular vector field.*

The following result from [39] is an extension of Theorem 4.

Theorem 10. *Let M be a submanifold of a Riemannian manifold* $\tilde{M}$*, which admits a torqued vector field* $\mathcal{T}$*. If the tangential component* $\mathcal{T}^T$ *of* $\mathcal{T}$ *is nonzero on M, then M is a rectifying submanifold (with respect to* $\mathcal{T}$*) if and only if* $\mathcal{T}^T$ *is torse-forming vector field on M whose conformal scalar is the restriction of the torqued function and whose generating form is the restriction of the torqued form of* $\mathcal{T}$ *on M.*

In [39], we also have the following results.

Theorem 11. *Let M be a submanifold of a Riemannian manifold* $\tilde{M}$ *endowed with a concircular vector field* $Z \neq 0$ *with* $Z^T \neq 0$ *on M. Then, M is a rectifying submanifold with respect to Z if and only if the tangential component* Z^T *of Z is a concircular vector field with the concircular function given by the restriction of the concircular function of Z on M.*

The following result is an immediate consequence of Theorem 11.

Corollary 2. *Let M be a submanifold of a Riemannian manifold* $\tilde{M}$ *endowed with a concurrent vector field* $Z \neq 0$ *such that* $Z^T \neq 0$ *on M. Then, M is a rectifying submanifold with respect to Z if and only if the tangential component* Z^T *of Z is a concurrent vector field on M.*

Moreover, from Theorem 11, we also have the following.

Proposition 3. *Let* $\tilde{M}$ *be a Riemannian m-manifold endowed with a concircular vector field Z. If M is a rectifying submanifold of* $\tilde{M}$ *with respect to Z, then we have:*

(1) Z^N *is of constant length* $\neq 0$.
(2) *The concircular function* φ *of* Z^T *is given by* $\varphi = Z^T(\ln \rho)$, *where* $\rho = |Z^T|$.

9. Euclidean Submanifolds with x^T as Potential Fields

A smooth vector field ξ on a Riemannian manifold (M, g) is said to define a Ricci soliton if it satisfies:

$$\frac{1}{2}\mathcal{L}_\xi g + Ric = \lambda g, \tag{39}$$

where $\mathcal{L}_\xi g$ is the Lie-derivative of the metric tensor g with respect to ξ, Ric is the Ricci tensor of (M, g) and λ is a constant (cf. for instance [46–48]). We shall denote a Ricci soliton by (M, g, ξ, λ).

A Ricci soliton (M, g, ξ, λ) is called shrinking, steady or expanding according to $\lambda > 0$, $\lambda = 0$, or $\lambda < 0$, respectively. A trivial Ricci soliton is one for which ξ is zero or Killing, in which case the metric is Einstein.

A Ricci soliton (M, g, ξ, λ) is called a gradient Ricci soliton if its potential field ξ is the gradient of some smooth function f on M.

For a gradient Ricci soliton, the soliton equation can be expressed as:

$$Ric_f = \lambda g, \tag{40}$$

where

$$Ric_f := Ric + Hess(f) \tag{41}$$

is known as the Bakry–Émery curvature, where $Hess(f)$ denotes the Hessian of f. Hence, a gradient Ricci soliton has constant Bakry–Émery curvature; a similar role as an Einstein manifold.

Compact Ricci solitons are the fixed points of the Ricci flow:

$$\frac{\partial g(t)}{\partial t} = -2Ric(g(t)) \tag{42}$$

projected from the space of metrics onto its quotient modulo diffeomorphisms and scalings and often arise as blow-up limits for the Ricci flow on compact manifolds. Further, Ricci solitons model the formation of singularities in the Ricci flow, and they correspond to self-similar solutions (cf. [47]).

During the last two decades, the geometry of Ricci solitons has been the focus of attention of many mathematicians. In particular, it has become more important after Grigory Perelman [48] applied Ricci solitons to solve the long-standing Poincaré conjecture posed in 1904. G. Perelman observed in [48] that the Ricci solitons on compact simply connected Riemannian manifolds are gradient Ricci solitons as solutions of Ricci flow.

The next result from ([31], Theorem 5.1) classifies Ricci solitons with concircular potential field.

Theorem 12. *A Ricci soliton (M, g, v, λ) on a Riemannian n-manifold (M, g) with $n \geq 3$ has concircular potential field v if and only if the following three conditions hold:*

(a) *The function φ in (31) is a nonzero constant, say b;*
(b) $\lambda = b$;
(c) *M is an open portion of a warped product manifold $I \times_{bs+c} F$, where I is an open interval with arc-length s, c is a constant and F is an Einstein $(n-1)$-manifold whose Ricci tensor satisfies*

$$Ric_F = (n-2)b^2 g_F,$$

where g_F is the metric tensor of F.

By combining Theorem 12 with some results from [31], we have the following .

Corollary 3. *The only Riemannian manifold of constant sectional curvature admitting a Ricci soliton with concircular potential field is a Euclidean space [31].*

Now, we present results on Ricci solitons of Euclidean hypersurfaces such that the potential field ξ is the tangential components $\mathbf{x}^T$ of the position vector field of the hypersurfaces.

For Ricci solitons on a Euclidean submanifold with the potential field given by $\mathbf{x}^T$, we have the following result from ([49], Theorem 4.1, p. 6).

Theorem 13. *Let (M, g, ξ, λ) be a Ricci soliton on a Euclidean submanifold M of $\mathbb{E}^m$. If the potential field ξ is the tangential component $\mathbf{x}^T$ of the position vector field of M, then the Ricci tensor of (M, g) satisfies:*

$$Ric(X, Y) = (\lambda - 1) \langle X, Y \rangle - \langle h(X, Y), \mathbf{x}^N \rangle \tag{43}$$

for any X, Y tangent to M.

Let ξ be a normal vector field of a Riemannian submanifold M. Then, M is called ζ-umbilical if its shape operator satisfies $A_\zeta = \varphi I$, where φ is a function on M and I is the identity map.

The following are some simple applications of Theorem 13.

Corollary 4. *A Ricci soliton $(M, g, \mathbf{x}^T, \lambda)$ on a Euclidean submanifold M is trivial if and only if M is $\mathbf{x}^\perp$-umbilical.*

Corollary 5. *Every Ricci soliton $(M, g, \mathbf{x}^T, \lambda)$ on a totally umbilical submanifold M of $\mathbb{E}^m$ is a trivial Ricci soliton.*

Corollary 6. *If $(M, g, \mathbf{x}^T, \lambda)$ is a Ricci soliton on a minimal submanifold M in $\mathbb{E}^m$, then M has constant scalar curvature given by $\frac{1}{2}n(\lambda - 1)$ with $n = \dim M$.*

Corollary 7. *Every Ricci soliton $(M, g, \mathbf{x}^T, \lambda)$ on a Euclidean submanifold M is a gradient Ricci soliton with potential function $\varphi = \frac{1}{2}\tilde{g}(\mathbf{x}, \mathbf{x})$.*

The next result was also obtained in ([49], Proposition 4.1, p. 6).

Theorem 14. *If (M, g, ξ, λ) is a Ricci soliton on a hypersurface of M of $\mathbb{E}^{n+1}$ whose potential field ξ is $\mathbf{x}^T$, then M has at most two distinct principal curvatures given by:*

$$\kappa_1, \kappa_2 = \frac{n\alpha + \rho \pm \sqrt{(n\alpha + \rho)^2 + 4 - 4\lambda}}{2}, \tag{44}$$

where α is the mean curvature and ρ is the support function of M, i.e., $\rho = \langle \mathbf{x}, N \rangle$ and $H = \alpha N$ with N being a unit normal vector field.

The following result from ([50], Theorem 4.2) classifies the Ricci soliton of Euclidean hypersurfaces with the potential field given by $\mathbf{x}^T$ (see also [51,52]).

Theorem 15. *Let $(M, g, \mathbf{x}^T, \lambda)$ be a Ricci soliton on a hypersurface of M of $\mathbb{E}^{n+1}$. Then, M is one of the following hypersurfaces of $\mathbb{E}^{n+1}$:*

(1) *A hyperplane through the origin zero.*
(2) *A hypersphere centered at the origin.*
(3) *An open part of a flat hypersurface generated by lines through the origin zero;*
(4) *An open part of a circular hypercylinder $S^1(r) \times \mathbb{E}^{n-1}$, $r > 0$;*
(5) *An open part of a spherical hypercylinder $S^k(\sqrt{k-1}) \times \mathbb{E}^{n-k}$, $2 \le k \le n-1$,*

where $n = \dim M$.

10. Interactions between Torqued Vector Fields and Ricci Solitons

In this section, we present some interactions between torqued vector fields and Ricci solitons on Riemannian manifolds from [40].

First, we recall the following definition.

Definition 4. *The twisted product $B \times_f F$ of two Riemannian manifolds (B, g_B) and (F, g_F) is the product manifold $B \times F$ equipped with the metric:*

$$g = g_B + f^2 g_F, \tag{45}$$

where f is a positive function on $B \times F$, which is called the twisting function. In particular, if the function f in (45) depends only B, then it is called a warped product, and the function f is called the warping function.

The following result from ([40], Theorem 2.1, p. 241) completely determined those Riemannian manifolds admitting torqued vector fields.

Theorem 16. *If a Riemannian manifold M admits a torqued vector field $\mathcal{T}$, then M is locally a twisted product $I \times_f F$ such that $\mathcal{T}$ is always tangent to I, where I is an open interval. Conversely, for each twisted product $I \times_f F$, there exists a torqued vector field $\mathcal{T}$ such that $\mathcal{T}$ is always tangent to I.*

In view of Theorem 16, we made in [40] the following.

Definition 5. *A torqued vector field $\mathcal{T}$ is said to be associated with a twisted product $I \times_f F$ if $\mathcal{T}$ is always tangent to I.*

We have the following result from [40].

Theorem 17. *Every torqued vector field $\mathcal{T}$ associated with a twisted product $I \times_f F$ is of the form:*

$$\mathcal{T} = \mu f \frac{\partial}{\partial s}, \tag{46}$$

where s is an arc-length parameter of I, μ is a nonzero function on F and f is the twisting function.

Theorem 18. *A torqued vector field $\mathcal{T}$ on a Riemannian manifold M is a Killing vector field if and only if $\mathcal{T}$ is a recurrent vector field that satisfies:*

$$\nabla_X \mathcal{T} = \alpha(X)\mathcal{T} \ \ and \ \ \alpha(\mathcal{T}) = 0, \tag{47}$$

where α is a one-form.

As an application of Theorem 17, we have the following classification of torqued vector fields on Einstein manifolds.

Theorem 19. *Every torqued vector field $\mathcal{T}$ on an Einstein manifold M is of the form:*

$$\mathcal{T} = \zeta Z, \tag{48}$$

where Z is a concircular vector field on M and ζ is a function satisfying $Z\zeta = 0$. Conversely, every vector field of the form (48) is a torqued vector field on M.

Another application of Theorem 17 is the following.

Corollary 8. *Up to constants, there exists at most one concircular vector field associated with a warped product $I \times_\eta F$.*

A Riemannian manifold (M, g) is called a quasi-Einstein manifold if its Ricci tensor Ric satisfies:

$$Ric = ag + b\alpha \otimes \alpha \tag{49}$$

for functions a, b, and one-form α.

A Riemannian manifold (M, g) is called a generalized quasi-Einstein [53] (resp., mixed quasi-Einstein [54] or nearly quasi-Einstein [55]) manifold if its Ricci tensor satisfies:

$$\begin{aligned} &Ric = ag + b\alpha \otimes \alpha + c\beta \otimes \beta, \\ &(\text{resp., } Ric = ag + b\alpha \otimes \beta + c\beta \otimes \alpha \ \text{ or } \ Ric = ag + bE) \end{aligned} \tag{50}$$

where a, b, c are functions, α, β are one-forms and E is a non-vanishing symmetric $(0,2)$-tensor on M.

In [40], we made the following definition.

Definition 6. *A pseudo-Riemannian manifold is called almost quasi-Einstein if its Ricci tensor satisfies:*

$$Ric = ag + b(\beta \otimes \gamma + \gamma \otimes \beta) \tag{51}$$

for some functions a, b and one-forms β and γ.

For Ricci solitons with torqued potential field, we have the following result from [40].

Theorem 20. *If the potential field of a Ricci soliton $(M, g, \mathcal{T}, \lambda)$ is a torqued vector field $\mathcal{T}$, then (M, g) is an almost quasi-Einstein manifold.*

The following result from [40] provides a very simple characterization for a Ricci soliton with torqued potential field to be trivial.

Theorem 21. *A Ricci soliton $(M, g, \mathcal{T}, \lambda)$ with torqued potential field $\mathcal{T}$ is trivial if and only if $\mathcal{T}$ is a concircular vector field.*

In view of Theorem 16, we made the following.

Definition 7. *For a twisted product $I \times_f F$, the torqued vector field $f\partial/\partial s$ is called the canonical torqued vector field of $I \times_f F$, where s is an arc-length parameter on I.*

We denote the canonical vector field $f\partial/\partial s$ by $\mathcal{T}^f_{ca}$.

Recall from Theorem 16 that if a Riemannian manifold M admits a torqued vector field, then it is locally a twisted product $I \times_f F$, where F is a Riemannian $(n-1)$-manifold and f is the twisting function. In [40], we proved the following.

Theorem 22. *If $(I \times_f F, g, \mathcal{T}^f_{ca}, \lambda)$ is a Ricci soliton with the canonical torqued vector field $\mathcal{T}^f_{ca}$ as its potential field, then we have:*

(a) *$\mathcal{T}^f_{ca}$ is a concircular vector field and*
(b) *$(I \times_f F, g)$ is an Einstein manifold.*

Remark 7. *Ricci solitons (M, g, Z, λ) with concircular potential field Z have been completely determined in ([31], Theorem 5.1).*

Remark 8. *If the potential field of the Ricci soliton defined on $(I \times_f F, g)$ in Theorem 16 is an arbitrary torqued vector field $\mathcal{T}$ associated with $I \times_f F$, then it follows from Theorem 17 that $\mathcal{T} = \mu f\partial/\partial s$ for some function μ defined on F. In this case, we may consider the twisted product $I \times_{\tilde{f}} \tilde{F}$ instead, where $\tilde{f} = \mu f$ and $\tilde{F}$ is the manifold F with metric $\tilde{g}_F = \mu^{-2} g_F$. Then, $(I \times_{\tilde{f}} \tilde{F}, \tilde{g}, \mathcal{T}, \lambda)$ with $\tilde{g} = ds^2 + \tilde{f}^2 \tilde{g}_F$ is a Ricci soliton whose potential field $\mathcal{T}$ is the canonical torqued vector field $\mathcal{T}^f_{ca}$ of $I \times_{\tilde{f}} \tilde{F}$.*

An important application of Theorem 22 is the following.

Corollary 9. *Let $(I \times_f F, g, \mathcal{T}^f_{ca}, \lambda)$ be steady Ricci solitons with the canonical torqued vector field $\mathcal{T}^f_{ca}$ as its potential field. If $\dim F \geq 2$, then we have:*

(a) *$\mathcal{T}^f_{ca}$ is a parallel vector field,*
(b) *f is a constant, say c,*

(c) $(I \times_c F, g)$ *is a Ricci-flat manifold and*
(d) F *is also Ricci-flat.*

11. Conclusions

The position vector field $\mathbf{x}$ is the most elemantary and natural object on a Euclidean submanifold. Similarly, the tangential component $\mathbf{x}^T$ of the position vector field is the most natural vector field tangent to the sumanifold. From the results we mentioned above, we conclude that the tangential component $\mathbf{x}^T$ of the position vector field of the Euclidean submanifold is the most important vector field naturally associated with the Euclidean submanifold. The author believes that many further important properties of $\mathbf{x}^T$ can be proved.

Conflicts of Interest: The authors declare no conflict of interest.

References

1. Chen, B.-Y. Topics in differential geometry associated with position vector fields on Euclidean submanifolds. *Arab J. Math. Sci.* **2017**, *23*, 1–17.
2. Chen, B.-Y. *Geometry of Submanifolds*; Marcel Dekker: New York, NY, USA, 1973.
3. Chen, B.-Y. *Riemannian Submanifolds, Handbook of Differential Geom*; Elsevier: Amsterdam, The Netherland, 2000; Volume 1, pp. 187–418.
4. Chen, B.-Y. *Pseudo-Riemannian Geometry, δ-Invariants and Applications*; World Scientific: Singapore, 2011.
5. Chen, B.-Y. *Differential Geometry of Warped Product Manifolds and Submanifolds*; World Scientific: Singapore, 2017.
6. O'Neill, B. *Semi-Riemannian Geometry with Applications to Relativity*; Academic Press: New York, NY, USA, 1983.
7. Chen, B.-Y. Differential geometry of rectifying submanifolds. *Int. Electron. J. Geom.* **2016**, *9*, 1–8.
8. Chen, B.-Y. Geometry of position functions of Riemannian submanifolds in pseudo-Euclidean space. *J. Geom.* **2002**, *74*, 61–77.
9. Chen, B.-Y. Constant-ratio hypersurfaces. *Soochow J. Math.* **2001**, *27*, 353–362.
10. Chen, B.-Y. Constant-ratio space-like submanifolds in pseudo-Euclidean space. *Houst. J. Math.* **2003**, *29*, 281–294.
11. Thompson, D. *On Growth and Form*; Cambridge University Press: Cambridge, UK, 1942.
12. Fu, Y.; Munteanu, M.I. Generalized constant ratio surfaces in $\mathbb{E}^3$. *Bull. Braz. Math. Soc. (N.S.)* **2014**, *45*, 73–90.
13. Haesen, S.; Nistor, A.I.; Verstraelen, L. On growth and form and geometry I. *Kragujevac J. Math.* **2012**, *36*, 5–25.
14. Benson, D.C. Motion in a central force field with drag or tangential propulsion. *SIAM J. Appl. Math.* **1982**, *42*, 738–750.
15. Lamb, H. *Dynamics*; Cambridge University Press: London, UK, 1923.
16. Nauenberg, M. Newton's early computational method for dynamics. *Arch. Hist. Exact Sci.* **1994**, *46*, 221–252.
17. Newton, I. *Principia*; Motte's Translation Revised; University of California: Berkeley, CA, USA, 1947.
18. Chen, B.-Y. More on convolution of Riemannian manifolds. *Beiträge Algebra Geom.* **2003**, *44*, 9–24.
19. Chen, B.-Y. Convolution of Riemannian manifolds and its applications. *Bull. Aust. Math. Soc.* **2002**, *66*, 177–191.
20. Chen, B.-Y. When does the position vector of a space curve always lie in its rectifying plane? *Am. Math. Mon.* **2003**, *110*, 147–152.
21. Chen, B.-Y. Rectifying curves and geodesics on a cone in the Euclidean 3-space. *Tamkang J. Math.* **2017**, *48*, 209–214.
22. Chen, B.Y.; Dillen, F. Rectifying curves as centrodes and extremal curves. *Bull. Inst. Math. Acad. Sin.* **2005**, *33*, 77–90.
23. Deshmukh, S.; Chen, B.-Y.; Alshammari, S.H. On rectifying curves in Euclidean 3-space. *Turkish J. Math.* **2018**, *42*, doi:10.3906/mat-1701-52.
24. Deshmukh, S.; Chen, B.-Y.; Turki, N.B. Unpublished work, 2017.
25. Chen, B.-Y.; Yano, K. On submanifolds of submanifolds of a Riemannian manifold. *J. Math. Soc. Jpn.* **1971**, *23*, 548–554.
26. Mihai, I.; Rosca, R.; Verstraelen, L. Some Aspects of the Differential Geometry of Vector Fields. On Skew Symmetric Killing and Conformal Vector Fields, and Their Relations to Various Geometrical Structures; Group of Exact Sciences, 1996; Volume 2. Available online: http://ci.nii.ac.jp/naid/10010836794/ (accessed on 28 August 2017).

27. Schouten, J.A. *Ricci-Calculus*, 2nd ed.; Springer: Berlin, Germany, 1954.
28. Yano, K.; Chen, B.-Y. On the concurrent vector fields of immersed manifolds. *Kodai Math. Sem. Rep.* **1971**, *23*, 343–350.
29. Chen, B.-Y. Addendum to: Differential geometry of rectifying submanifolds. *Int. Electron. J. Geom.* **2017**, *10*, 81–82.
30. Chen, B.Y.; Oh, Y.M. Classification of rectifying space-like submanifolds in pseudo-Euclidean spaces. *Int. Electron. J. Geom.* **2017**, *10*, 86–95.
31. Chen, B.-Y. Some results on concircular vector fields and their applications to Ricci solitons. *Bull. Korean Math. Soc.* **2015**, *52*, 1535–1547.
32. Yano, K. Concircular geometry. I. Concircular transformations. *Proc. Imp. Acad. Tokyo* **1940**, *16*, 195–200.
33. Chen, B.-Y.; Wei, S.W. Differential geometry of concircular submanifolds of Euclidean spaces. *Serdica Math. J.* **2017**, *43*, 36–48.
34. Chen, B.-Y. A simple characterization of generalized Robertson-Walker spacetimes. *Gen. Relativ. Gravit.* **2014**, *46*, 1833.
35. Mantica, C.A.; Molinari, L.G. Generalized Robertson-Walker spacetimes—A survey. *Int. J. Geom. Methods Mod. Phys.* **2017**, *14*, 1730001.
36. Chen, B.-Y. Concircular vector fields and pseudo-Kaehler manifold. *Kragujevac J. Math.* **2016**, *40*, 7–14.
37. Yano, K. On torse forming direction in a Riemannian space. *Proc. Imp. Acad. Tokyo* **1944**, *20*, 340–345.
38. Mihai, A.; Mihai, I. Torse forming vector fields and exterior concurrent vector fields on Riemannian manifolds and applications. *J. Geom. Phys.* **2013**, *73*, 200–208.
39. Chen, B.-Y. Rectifying submanifolds of Riemannian manifolds and torqued vector fields. *Kragujevac J. Math.* **2017**, *41*, 93–103.
40. Chen, B.-Y. Classification of torqued vector fields and its applications to Ricci solitons. *Kragujevac J. Math.* **2017**, *41*, 239–250.
41. Alías, L.J.; Romero, A.; Sánchez, M. Uniqueness of complete spacelike hypersurfaces of constant mean curvature in generalized Robertson-Walker spacetimes. *Gen. Relativ. Gravit.* **1995**, *27*, 71–84.
42. Chen, B.-Y. *Geometry of Submanifolds and Applications*; Science University of Tokyo: Tokyo, Japan, 1981.
43. Mantica, C.A.; Molinari, L.G. Twisted Lorentzian manifolds, a characterization with torse-forming time-like unit vectors. *Gen. Relativ. Gravit.* **2017**, *49*, 51.
44. Chen, B.-Y.; Verstraelen, L. A link between torse-forming vector fields and rotational hypersurfaces. *Int. J. Geom. Methods Mod. Phys.* **2017**, *14*, 1750177.
45. Chen, B.-Y.; Yano, K. Special conformally flat spaces and canal hypersurfaces. *Tohoku Math. J.* **1973**, *25*, 177–184.
46. Chen, B.-Y.; Deshmukh, S. Geometry of compact shrinking Ricci solitons. *Balkan J. Geom. Appl.* **2014**, *19*, 13–21.
47. Morgan, J.; Tian, J. *Ricci Flow and the Poincaré Conjecture*; Clay Mathematics Monographs; The American Mathematical Societ: Cambridge, MA, USA, 2014; Volume 5.
48. Perelman, G. The entropy formula for the Ricci flow and its geometric applications. *arXiv* **2002**, arXiv:math/0211159.
49. Chen, B.-Y.; Deshmukh, S. Ricci solitons and concurrent vector fields. *Balkan J. Geom. Appl.* **2015**, *20*, 14–25.
50. Chen, B.-Y.; Deshmukh, S. Classification of Ricci solitons on Euclidean hypersurfaces. *Int. J. Math.* **2014**, *25*, 1450104.
51. Chen, B.-Y. Ricci solitons on Riemannian submanifolds. In Riemannian Geometry and Applications; University of Bucharest Press: Bucharest, Romania, 2014; pp. 30–45.
52. Chen, B.-Y. A survey on Ricci solitons on Riemannian submanifolds. In *Recent Advances in the Geometry of Submanifolds*; American Mathematical Society: Providence, RI, USA, 2016; Volume 674, pp. 27–39.
53. De, U.C.; Ghosh, G.C. On generalized quasi-Einstein manifolds. *Kyungpook Math. J.* **2004**, *44*, 607–615.
54. Mallick, S.; De, U.C. On mixed quasi-Einstein manifolds. *Ann. Univ. Sci. Budapest. Eötvös Sect. Math.* **2014**, *57*, 59–73.
55. De, U.C.; Gazi, A.K. On nearly quasi Einstein manifolds. *Novi Sad J. Math.* **2008**, *38*, 115–121.

mathematics

MDPI

Article

A New Proof of a Conjecture on Nonpositive Ricci Curved Compact Kähler–Einstein Surfaces

Zhuang-Dan Daniel Guan

Department of Mathematics, The University of California at Riverside, Riverside, CA 92521, USA; zguan_02@yahoo.com

Received: 26 December 2017; Accepted: 26 January 2018; Published: 7 February 2018

Abstract: In an earlier paper, we gave a proof of the conjecture of the pinching of the bisectional curvature mentioned in those two papers of Hong et al. of 1988 and 2011. Moreover, we proved that any compact Kähler–Einstein surface M is a quotient of the complex two-dimensional unit ball or the complex two-dimensional plane if (1) M has a nonpositive Einstein constant, and (2) at each point, the average holomorphic sectional curvature is closer to the minimal than to the maximal. Following Siu and Yang, we used a minimal holomorphic sectional curvature direction argument, which made it easier for the experts in this direction to understand our proof. On this note, we use a maximal holomorphic sectional curvature direction argument, which is shorter and easier for the readers who are new in this direction.

Keywords: Kähler–Einstein metrics; compact complex surfaces; pinching of the curvatures

1. Introduction

In [1], the authors conjectured that any compact Kähler–Einstein surface with negative bisectional curvature is a quotient of the complex two-dimensional unit ball. They proved that there is a number $a \in (1/3, 2/3)$ such that if at every point P, $K_{av} - K_{min} \leq a(K_{max} - K_{min})$, then M is a quotient of the complex ball. Here, K_{min} (K_{max}, K_{av}) is the minimal (maximal, average) of the holomorphic sectional curvature. The number a they obtained is $a < \frac{2}{3[1+\sqrt{6/11}]}$ (almost 0.38; see [2], p. 398). In [3], Yi Hong pointed out that this is also true if $a \leq \frac{2}{3[1+\sqrt{1/6}]} < 0.476$. (For this part, this is due to Professor Hong. One note that he was the first author there.) We also observed in Theorem 2 that if $a \leq \frac{1}{2}$, then there is a ball-like point P. That is, at P, $K_{max} = K_{min}$. We notice that $\sqrt{1/6} > 1/3$. Therefore, we conjectured in [3] that M is a quotient of the complex ball if $a = \frac{1}{2}$. In general, we believe that we might not obtain a quotient of the complex ball if $a > \frac{1}{2}$. In [2,4], the author used a different method and proved that a can be $(3 + \frac{4\sqrt{3}}{3})/11$ (almost 0.48 according to [5], p. 2628, just before Theorem 1.2); see [4] (p. 669) or [2] (p. 398). In [5], the authors improved the constant to $a < \frac{1}{2}$, which gave a proof of a weaker version of the conjecture.

In [6], we proved the following:

Proposition 1. Let M be a connected compact Kähler–Einstein surface with nonpositive scalar curvature; if we have

$$K_{av} - K_{min} \leq \frac{1}{2}[K_{max} - K_{min}]$$

at every point, then M is a compact quotient of either the complex two-dimensional unit ball or the complex two-dimensional complex plane.

For important mathematics work, it is common practice to give different and (possibly) simpler proofs (for certain experts and readers). For examples, see [7–14] and so forth.

This note is for the experts who are new in this direction. In the second section, we review the basic material from [1] with an emphasis on the maximal holomorphic sectional direction instead of the minimal holomorphic sectional direction in [1,3,5,6]. We prove the existence of the ball-like points, as we did in the second section in [6], by using a different but similar function. In the third section, we again use Hong-Cang Yang's function and a different but similar calculation with respect to the maximal direction instead of the minimal direction. We give some detailed calculations in the Appendix as the last section of this paper.

2. Existence of Ball-Like Points

Here, we repeat the argument in the proof of our Proposition 1 given in [6] by using a different but similar argument:

Proposition 1 (cf. [3] pp. 597–599; [6] Proposition 1) Suppose that

$$K_{av} - K_{min} \leq \frac{1}{2}[K_{max} - K_{min}]$$

for every point on the compact Kähler–Einstein surface with nonpositive Ricci curvatures. There is at least one ball-like point.

Proof of Proposition 1. Throughout this section, as in [1,5,6], we assume that $\{e_1, e_2\}$ is a unitary basis at a given point P with

$$R_{1\bar{1}1\bar{1}} = R_{2\bar{2}2\bar{2}} = K_{min},\ R_{1\bar{1}1\bar{2}} = R_{2\bar{2}2\bar{1}} = 0$$
$$A = 2R_{1\bar{1}2\bar{2}} - R_{1\bar{1}1\bar{1}} \geq 0,\ B = |R_{1\bar{2}1\bar{2}}|$$

As in [1], we always have that $A \geq |B|$, and we assume that $B \geq 0$. This also implies, if the sectional curvatures have a $1/4$ pinching, that is, the section curvature is inside an interval $[-\frac{1}{4}a(P), -a(P)]$ at every point P for a nonnegative function $a(P)$, that M is covered by a ball. This was pointed out in [5]. This is because if we let $a(P) = -R_{1\bar{1}1\bar{1}}$, $e_i = X_i + \sqrt{-1}Y_i$, then at least one of $R(X_1, X_2, X_1, X_2)$ and $R(X_1, Y_2, X_1, Y_2)$ is greater than or equal to $-\frac{1}{4}a(P)$. The same argument works for the higher-dimensional case. Our proposition is a kind of the generalization of the $1/4$ pinching.

If P is not a ball-like point, according to [1], we can do the above for a neighborhood $U(P)$ of P whenever $A > B$ (case 1 in [1], p. 475). In [6], we took great effort in handling the case in which $A = B$. We write

$$\alpha = e_1 = \sum a_i \partial_i,\ \beta = e_2 = \sum b_i \partial_i$$

and

$$S_{1\bar{1}1\bar{1}} = R(e_1, \bar{e}_1, e_1, \bar{e}_1) = \sum R_{i\bar{j}k\bar{l}} a_i \bar{a}_j a_k \bar{a}_l$$

and so on. In particular, we have

$$S_{1\bar{1}1\bar{1}} = S_{2\bar{2}2\bar{2}} = K_{min},\ S_{1\bar{1}1\bar{2}} = S_{2\bar{2}2\bar{1}} = 0$$

According to [1], we have

$$K_{max} = K_{min} + \frac{1}{2}(A + B),\ K_{av} = K_{min} + \frac{1}{3}A$$

$$\frac{1}{3}[K_{max} - K_{min}] \leq K_{av} - K_{min} \leq \frac{2}{3}[K_{max} - K_{min}]$$

This also shows that A and B are independent of the choice of e_1 and e_2. Additionally, our condition in Proposition 1 is therefore the same as $A \leq 3B$.

In this section, we denote the maximal direction by e_{1*} and use $*$ in the notation of the corresponding terms' minimal direction case. We assume that P is not a ball-like point. Under our assumption, $B > 0$. According to ([1], p. 474), e_{1^*} could be $\frac{1}{\sqrt{2}}(e_1 + e_2)$. We could pick up $e_{2^*} = \frac{i}{\sqrt{2}}(e_1 - e_2)$. We have

$$A^* = 2R_{1^*\bar{1}^*2^*\bar{2}^*} - R_{1^*\bar{1}^*1^*\bar{1}^*} = -\frac{1}{2}(A + 3B)$$

$$B^* = R_{1^*\bar{2}^*1^*\bar{2}^*} = \frac{1}{2}(A - B)$$

In our case, we have $A^* + 3B^* = A - 3B \leq 0$, that is, $-A^* \geq 3B^*$. Moreover, from both the arguments in [1] (pp. 474–475), the choices of the directions of e_{1^*} are isolated on the projective holomorphic tangent space. These two cases are case 1: $A > B$; and case 2: $A = B$. In case 1, there is only one direction for the minimal holomorphic sectional curvature, and there is only one direction for the maximal holomorphic sectional curvature, because, by our assumption, $A \leq 3B$; that is, B is not zero at a nonball-like point. The second statement also follows from the argument in [1] by applying it to the maximal direction instead of the minimal direction. In case 2, there is a circle for the minimal direction, but there is a unique maximal direction. That is, one could always have a smooth frame of e_{1^*}. This might make the proof simpler. However, near the points with $B^* = A - B = 0$, we might still have difficulty to obtain a smooth frame nearby such that $B^* \geq 0$. Therefore, we only assume that $B^* \geq 0$ at P, but this is not necessary true nearby if $B^*(P) = 0$.

In [3,6], we let $\Phi_1 = \frac{|B|^2}{A^2} = \tau^2$. Here, we let $\Phi_1^* = \frac{|B^*|^2}{(A^*)^2} = (\tau^*)^2$ and $\tau^* = -\frac{|B^*|}{A^*} \geq 0$.

Our condition is the same as $\tau^* \leq 1/3$. If there is no ball-like point, there is a maximal point.

Now, $\tau^* = \frac{A-B}{A+3B} = \frac{1}{3}(-1 + \frac{4}{1+3\tau})$. The maximal of τ^* is just the minimal of τ.

The calculation of the Laplace of Φ_1 at a minimal point, which is not a ball-like point, and $A \neq B$ in [6] showed that $B^* = 0$.

A similar calculation of the Laplacian of Φ_1^* with $B^* \neq 0$ shows that

$$\Delta\Phi_1^* = 6A^*(\tau^*)^2((\tau^*)^2 - 1) + h^* \tag{1}$$

Here $\Delta\Phi_1^*$ has two general terms, just as for the formula for $\Delta\Phi_1$ in [6]; see the Appendix at the end of this paper. The first term is always nonnegative, as $\tau^* \leq \frac{1}{3} \leq 1$. The second term is a Hermitian form h^* to y^*. We can separate y^* into two groups: y_{2j}^* in one group and y_{1j}^* in the other. These two groups of variables are orthogonal to each other with respect to this Hermitian form. That is, $h^* = h_1^* + h_2^*$ with h_1^* (or h_2^*) only depends on the first (second) group of variables.

We need to check the nonnegativity for each of these.

For y_{11}^*, y_{12}^*, the corresponding matrix of h_2^* is

$$\begin{bmatrix} 2(9(\tau^*)^2 - 1)((\tau^*)^2 - 1) & 0 \\ 0 & 0 \end{bmatrix}$$

and the matrix for h_1^* of y_{21}^*, y_{22}^* is

$$\begin{bmatrix} 0 & 0 \\ 0 & 2(9(\tau^*)^2 - 1)((\tau^*)^2 - 1) \end{bmatrix}$$

When P is a critical point of Φ_1^*, the matrices on y^* are clearly semipositive. Therefore, if there is no ball-like point, then we have that at the maximal point of Φ_1^*, $\tau^* = 0$ or $A^* = 0$, because $\tau^* \leq \frac{1}{3}$.

If $A^* = 0$, then we have a ball-like point. Thus we are done.

On the other hand, if $\tau^* = 0$, we have $B^* = 0$ at P. Because P is a maximal point for τ^*, this implies that $B^* = 0$ on the whole manifold. In this case, we could always assume that $B^* \geq 0$.

According to [1] (p. 475, case 2), that is, when $A = B$, we have smooth coordinates with $K_{max} = R_{1\bar{1}1\bar{1}}$ (this fortunately always works when $A = B$. In general, the original argument might not always work, as one might not have $A = B$ always nearby. However, as [1], case 1 also works for the maximal direction instead of the minimal direction, this implies that under our condition, the directions for K_{max} are always isolated. Therefore, it might be better for one to choose K_{max} instead of K_{min} from the very beginning). Using this new coordinate, we can define the similar functions A^* and B^*. In general, $B^* = \frac{1}{2}(A - B)$ and $A^* = -\frac{1}{2}(A + 3B)$. In our case, $B^* = 0$ and $A^* = -2A$. Using this new coordinate, one can do the calculation for any of the functions in [1,2,4] (or [5]; see the next section), for which the set of ball-like points is the whole manifold. If one does not like Polombo's function Φ_α ([2], p. 418) with $\alpha = -\frac{8}{7}$ (e.g., [2], p. 417, Lemma), then one might simply use the function with $\alpha = -1$ (in [2,4]; not the vector we mentioned in this paper earlier), that is, the new function is proportional to $\Phi_2 = (3B - A)A$. In our case, this is just $2A^2$. We can apply $\Phi_2^{\frac{1}{3}}$. This is relatively easy; thus we leave it to the readers (or see Equation (4) in the generalization). Actually, this paragraph is not needed for the proofs of Corollary 1 and Lemma 1. Additionally, in this special case, the original frame in [1] works. Thus, one could simply apply [1].

One can also use the function in [1] (p. 477):

$$3\gamma_2 - \gamma_1^2 = \frac{1}{2}(A^2 + 3B^2)$$

We can also still use the argument in [1], case 1, in which the minimal vectors are no longer isolated but are points in a smooth circle bundle over the manifold, such that we could choose a smooth section instead.

This paragraph is also not needed in the following Corollary 1 and Lemma 1, as in these two propositions, we already have $A = 3B$. With $A = B$, one can readily obtain that $A = B = 0$.

If $A = 0$, $K_{max} = K_{min}$ and P is a ball-like point, we have a contradiction. Therefore, the set of ball-like points is not empty.

Q. E. D.

We observe that if $A = 3B$ at P, then Φ_1 achieves the minimal value at P and $A \neq B$ unless P is a ball-like point. That is, the first part of the proof of Proposition 1 goes through; thus, P must be a ball-like point. □

Corollary 1. Assuming the above, if $K_{av} - K_{min} = \frac{1}{2}(K_{max} - K_{min})$ at P, then P is a ball-like point.

Therefore, we have the following:

Lemma 1. If $K_{av} - K_{min} \leq \frac{1}{2}[K_{max} - K_{min}]$ on M, then we have $K_{av} - K_{min} < \frac{1}{2}[K_{max} - K_{min}]$ on $M - N$, where N is the subset of all the ball-like points.

Therefore, we can apply the argument of [5]. To do this, one needs the following Proposition 4 in [1]:

Proposition 2. *(cf. [1,3], Theorem 3.)* If $N \neq M$, then N is a real analytic subvariety and codim $N \geq 2$.

As in [1], Proposition 2 gives us a way to the conjecture by finding a superharmonic function on M, which was obtained by Hong-Cang Yang around 1992. In [1,3], the authors used $\Phi = 6B^2 - A^2$. In [2], Polombo used $(11A - 3B)(B - A) + 16AB$; see [2] (p. 417, Lemma. One might ask why we need another function but do not use our Φ_1; the answer is that by a power of Φ_1, we can only correct the Laplace by $|\nabla\Phi_1|^2$. However, this could only change the upper left coefficients of our matrices as it only provides $|x|^2$ terms. In the case of Φ_1, it does not work, as $\frac{T}{A} \neq 0$ but the coefficients of $|y_{12}|^2, |y_{21}|^2$ are zeros.

Therefore, we need another function, which was provided by Hong-Cang Yang.

Remark 1. *Whenever there is a bounded continuous nonnegative function f on M such that (1) $f(N) = 0$, (2) f is real analytic on $M - N$, and (3) $\Delta f \leq 0$ on $M - N$, then $f = 0$. Here N could be just a codimension 2 subset. This is true in general for extending continuous superharmonic functions over a codimension 2 subset; see [1,3,5]. Here, we wish to give our own reasons as to why this is true in these special cases. If we define $M_s = \{x \in M|_{\text{dist}(x,N) \geq s}\}$ and $h_s = \partial M_s$, then the measure of h_s is smaller than $O(s)$ when s tends to zero. Therefore,*

$$0 \geq \ln 2 \int_{M_{2\delta}} \Delta f \omega^n \geq \int_{\delta}^{2\delta} [\int_{M_s} \Delta f \omega^n] s^{-1} ds = \int_{\delta}^{2\delta} [\int_{h_s} \frac{\partial f}{\partial n} d\tau] s^{-1} ds$$

However, by applying an integration by parts to the single-variable integral, the last term is about $(\delta)^{-1} \int_{h_{2\delta}} (f - g) d\tau \to 0$, as f is bounded and $f - g$ tends to 0 near N, where g is the f value of the corresponding point on h_δ. For example, if $f = r^a$ with $a > 0$, then

$$\frac{\partial f}{\partial n} = a r^{a-1} = a s^{a-1}$$

and

$$\int_{h_s} \frac{\partial f}{\partial n} d\tau = O(s^a) \to 0$$

Therefore, $\Delta f = 0$ on $M - N$. Therefore f extends over N as a harmonic function. This implies that $f = 0$ on M.

Now, letting $f = (3B - A)^a$, which is natural after the proof of Proposition 1, we show in the next section that $\Delta f \leq 0$ for $a \leq \frac{1}{3}$ (see also a proof in [5]). Therefore, $A = 3B$ always. By Corollary 1, we have $A = B = 0$. This function is also related to the functions in [2] (p. 417) with $a_1 = a_3 = 0$. In [2], Polombo had to pick up functions with $a_1 = a_2$ to avoid a complication of the singularities; see [2] (p 406 and the first paragraph on p. 418); see also [4] (the last paragraph of p. 668). However, we completely resolve the difficulty in the next section.

3. Generalized Hong-Cang Yang's Function

We let $\Psi = 3B - A = -A^* - 3B^*$. Around 1992, Hong-Cang Yang considered $f = \Psi^{\frac{1}{3}}$. In [5], the authors had a formula for the Laplacian of Ψ. To apply the method to the maximal direction, we notice that the same formula holds. Moreover, if we let $\Psi_k = 3B + kA$, then $\Psi = \Psi_{-1}$ and we have the following:

Lemma 2. *(cf. [5], p. 2630, Equation (13).)* We have $A = \frac{3B^* - A^*}{2}$,

$$B = -\frac{A^* + B^*}{2}$$

and

$$\begin{aligned} \Delta(3B + kA) &= 3[\Psi_k R_{1\bar{1}2\bar{2}} - B(3A + kB)] \\ &+ \frac{3}{B} |\nabla(\text{Im} R_{1\bar{2}1\bar{2}})|^2 + 6[(B + kA) \sum |y|^2 + 2(A + kB) \text{Re} \sum y_{i1} \bar{y}_{i2}] \end{aligned}$$

In particular, we have

$$\begin{aligned} \Delta\Psi &= 3[\Psi R_{1^*\bar{1}^*2^*\bar{2}^*} + B^*(3A^* + B^*)] \\ &- \frac{3}{B^*} |\nabla(\text{Im} R_{1^*\bar{2}^*1^*\bar{2}^*})|^2 - 6(A^* + B^*) \sum |y^*_{i1} + y^*_{i2}|^2 \end{aligned}$$

It is clear that in the case of the maximal direction, we have to assume $B^* \neq 0$. That is, we still need to deal with the case in which $A = B$. This is because, in general, one cannot calculate the second derivatives of B^* even if we could obtain a smooth frame near the considered point. Therefore, we still need deal with the singularities, as we did in our earlier paper.

*To make things easier for us, in the rest of this section and in the next section (except for Remark 2), we use notation without * for the maximal direction instead of the minimal direction if there is no confusion.*

We let $z_i = \nabla_i \Psi$. Then

$$-z_1 = \nabla_1(3B + A) = \frac{3}{2}\nabla_1(R_{1\bar{2}1\bar{2}} + R_{2\bar{1}2\bar{1}} - 2R_{1\bar{1}1\bar{1}})$$

$$\begin{aligned}\sqrt{-1}\nabla_1(\mathrm{Im}R_{1\bar{2}1\bar{2}}) &= \frac{1}{2}\nabla_1(R_{1\bar{2}1\bar{2}} - R_{2\bar{1}2\bar{1}}) \\ &= -\frac{1}{3}z_1 - \nabla_1 R_{2\bar{1}2\bar{1}} + \nabla_1 R_{1\bar{1}1\bar{1}} \\ &= -\frac{1}{3}z_1 - \nabla_2 \bar{R}_{1\bar{1}1\bar{2}} - \nabla_2 R_{1\bar{1}1\bar{2}} \\ &= -\frac{1}{3}z_1 + (A+B)y_{22} + (B+A)y_{21}\end{aligned}$$

$$-z_2 = \nabla_2(3B + A) = \frac{3}{2}\nabla_2(R_{2\bar{1}2\bar{1}} + R_{1\bar{2}1\bar{2}} - 2R_{1\bar{1}1\bar{1}})$$

$$\begin{aligned}\sqrt{-1}\nabla_2(\mathrm{Im}R_{1\bar{2}1\bar{2}}) &= \frac{1}{2}\nabla_2(R_{1\bar{2}1\bar{2}} - R_{2\bar{1}2\bar{1}}) \\ &= \frac{1}{3}z_2 - \nabla_2 R_{1\bar{1}1\bar{1}} + \nabla_2 R_{1\bar{2}1\bar{2}} \\ &= \frac{1}{3}z_2 - \nabla_1 R_{2\bar{1}1\bar{1}} - \nabla_1 R_{1\bar{1}1\bar{2}} \\ &= \frac{1}{3}z_2 + (B+A)y_{12} + (A+B)y_{11}\end{aligned}$$

We can write the formula in Lemma 2 as

$$\begin{aligned}\Delta\Psi &= 3[\Psi R_{1\bar{1}2\bar{2}} + B(B+3A)] \\ &+ 3\frac{A+B}{B}\Psi\sum|y_{i1}+y_{i2}|^2 \\ &- 2\frac{A+B}{B}\mathrm{Re}[(y_{12}+y_{11})\bar{z}_2 - (y_{22}+y_{21})\bar{z}_1] - \sum\frac{1}{3B}|z|^2\end{aligned} \tag{2}$$

Similarly to what we have in the last section, we have two general terms: the first is negative as the constant term of z and y; the second is a Hermitian form on z and y. We can let $w_i = y_{i^*1} - y_{i^*2}$ with $i^* \neq i$. Then the second term is a sum of two Hermitian forms. One of these is on w_1, z_1, and the other is on w_2, z_2. We notice that the second term is also nonpositive on y (or nonpositive on w, if we assume that $z = 0$). We can modify the coefficient of $|z|^2$ (only) by taking the power of Ψ. More precisely, if we let $g = \Psi^a$, to make sure that $\Delta g < 0$, after taking out a factor $3\frac{A+B}{B}$, we need

$$\begin{vmatrix} \Psi & 1/3 \\ 1/3 & -\frac{1+3\Psi^{-1}(1-a)B}{9(A+B)} \end{vmatrix} \geq 0$$

That is,

$$A + 3B - 3(1-a)B - A - B = (3a-1)B \leq 0$$

We have $1 - 3a \geq 0$. Thus, $a \leq 1/3$.

Therefore, we have the following:

Lemma 3. $\Delta g < 0$ for $a \leq 1/3$ on $M - N$.

This is exactly the same as what was obtained in [5]. In fact, the number 1/6 was already used in [1–4] for those quadratic functions.

Thus, finally we have the following:

Theorem 1. If $K_{av} - K_{min} \leq \frac{1}{2}[K_{max} - K_{min}]$, then M has a constant holomorphic sectional curvature.

Remark 2. *The reason we did not come to this earlier was that there was a difficulty when $A = B$. In that case, the argument in [1] (p. 475, case 2) seems not to work. Polombo resolved the problem by using a function that is symmetric to $\lambda_1 = -\frac{A}{3}$ and $\lambda_2 = \frac{A-3B}{6}$ (see [2], p. 418, first paragraph and the end of p. 397). However, Hong-Cang Yang's function Ψ is only $-6\lambda_2$ and therefore is not symmetric after all. To overcome the difficulty, we let $\Omega = \{x \in M|_{A=B}\}$. Then according to [1], all our calculations are sound on $M - \Omega$, because $N \subset \Omega$. In [5] (p. 2632), there was a suggestion to prove that* codim $\Omega \leq 2$, *although it was not very well explained. Then, everything went through. The relation was that if we use the argument in [1] (p. 475, case 2), using the maximal instead of the minimal, we let $B_1 = |R_{1\bar{2}1\bar{2}}|$ (or B^* as we did earlier); then $2B_1 = A - B$. That is, $\Omega = \{x \in M|_{B_1=0}\}$. The argument goes as follows:*

Case 1: If Ω is a closed region, we have

$$\begin{aligned} 0 &\geq \int_{M-\Omega} \Delta g \\ &= a \int_{-\partial\Omega} \Psi^{a-1} \frac{\partial(-A_1 - 3B_1)}{\partial n} \\ &\geq a \int_{-\partial\Omega} (2A)^{a-1} \frac{\partial(-A_1)}{\partial n} \\ &= -\int_{\Omega} \Delta F_1 \geq 0 \end{aligned}$$

where F_1 can be chosen from one of the functions in [2] that satisfies the symmetric condition on M, for example, a power of Φ_2 as in the proof of Proposition 1, or one of our functions with a calculation using the new smooth coordinate in [1] (p. 475) with $R_{1\bar{1}1\bar{1}} = K_{max}$ (see Equation (4) in the next section). In fact, A_1 itself is proportional to λ_2 in [2] and is symmetric in the sense of Polombo. On Ω, F_1 is just our g, as $B_1 = 0$. We notice that there is a sign difference for the Laplace operator in [2]. Again, on Ω, because $A = B$ on a neighborhood, the set of minimal directions is a S^1 bundle over Ω; therefore, one might choose a smooth section of it locally such that the calculation of [1] still works in our case. That is, one could simply choose F_1 to be g.

Case 2: If Ω is a hypersurface, the same argument goes through, except that $\int_{\partial(M-\Omega)} (A)^{a-1} \frac{\partial A}{\partial n} = 0$, because $A \neq 0$ outside a codimension 1 subset, and on $\Omega_1 = \{x \in \Omega|_{A \neq 0}\}$, the integral is integrated from both sides.

Therefore, Ω is a subset of codimension 2, and we can apply Remark 1. By the calculation in Remark 1, we see that g is harmonic on $M - \Omega$. Now, by Lemma 2, this implies that $B(B - 3A) = 0$, and hence $A = B = 0$ by our assumptions.

4. The Generalization

In fact, in the first section of [1], the authors did not require any negativity. We also see that in our second section, we also do not need any negativity, except when we apply the formula in Lemma 2 in Section 3.

In the first section of [1], they also considered the coordinate in which $R_{1\bar{1}1\bar{1}}$ achieves the maximal instead of the minimal. By using the maximal direction, it is much easier to see that the constant term in the Laplacian is negative. We only need to check the following:

$$
\begin{aligned}
C &= R_{1\bar{1}2\bar{2}} \\
&= k - R_{1\bar{1}1\bar{1}} \\
&= k/2 - (K_{\max} - k/2) \\
&= k/2 - (K_{\max} - K_{min}) \\
&= k/2 + A/3 \leq 0
\end{aligned} \tag{3}
$$

One might compare this with [6] to see the advantage of this new method.

Now, with $C \leq 0$, we could also easily cover the arguments both at the end of the proof of Proposition 1 and in Remark 2 in the case of $B = 0$ (using the maximal direction). Similarly to the calculation in Section 2, we obtain the following (see also [15] (p. 27) for a good calculation of this Laplacian at a maximal direction for any complex dimension):

$$\Delta R_{1\bar{1}1\bar{1}} = -AC + B^2 = -AC \leq 0$$

We also have

$$\nabla R_{1\bar{1}1\bar{2}} = -A\nabla a_2 - B\nabla \bar{a}_2 = -A\nabla a_2,$$

$$\Delta S_{1\bar{1}1\bar{1}} = -2A\sum |y|^2 - AC$$

$$\nabla_1 A = -3\nabla S_{1\bar{1}1\bar{1}} = -3Ay_{21},$$

$$\nabla_2 A = 3Ay_{12}$$

$$\nabla_{\bar{1}} R_{1\bar{2}1\bar{2}} = -A\bar{y}_{22} = 0,$$

$$\nabla_2 R_{1\bar{2}1\bar{2}} = Ay_{11} = 0$$

$$
\begin{aligned}
\Delta(|A|^a) = 3a|A|^{a-1}\Delta S_{1\bar{1}1\bar{1}} + a(a-1)|A|^{a-2}|\nabla A|^2 \\
&= 3a \times (-A)^{a-1}(-2A\sum |y|^2 - AC) \\
&+ 9a(a-1)(-A)^a \sum |y|^2 \\
&= 3a(-A)^a[(2 - 3(a-1))\sum |y|^2 + C]
\end{aligned} \tag{4}
$$

This is nonpositive when $a \leq 1/3$. This is the same as in Lemma 3 and for that in [5].

Therefore, we conclude the general case. One might conjecture that our theorem is also true in the higher-dimensional cases.

Remark 3. We note that this generalization essentially covers the results in [2,4] for the Kähler–Einstein case (see [2], p. 398, Corollary; see also [16], p. 415, Proposition 2 for W^+ for a Kähler surface). One might ask whether our result could be generalized to the Riemannian manifolds with closed half Weyl curvature tensors. This is out of the scope of this paper, although a similar result is true, that is, if $\lambda_2 \leq 0$ at every point. To make the relation between this paper and [2,4] clearer to the readers, we mention that any one of the half Weyl tensors is harmonic if and only if it is closed, because the tensor is dual to either itself or the negative of itself. Remark (i) in [2] (p. 397) states that if M is Riemannian–Einsteinian, the second Bianchi identity states that the half Weyl tensors are closed (see also [16], p. 408, Equation (9) and p. 411, Remark 1).

5. Appendix

Here, we repeat the argument in the proof of Proposition 1 in [6] by using a different but similar argument.

Throughout this Appendix, as in [1,5,6], we assume that $\{e_1, e_2\}$ is a unitary basis at a given point P with

$$R_{1\bar{1}1\bar{1}} = R_{2\bar{2}2\bar{2}} = K_{min}$$

or K_{max}.

$$R_{1\bar{1}1\bar{2}} = R_{2\bar{2}2\bar{1}} = 0$$

$$A = 2R_{1\bar{1}2\bar{2}} - R_{1\bar{1}1\bar{1}} \geq 0$$

or ≤ 0 in the maximal direction case.

$$B = |R_{1\bar{2}1\bar{2}}|$$

As in [1], we always have that $A \geq |B|$ or $-A \geq |B|$. We assume that $B \geq 0$.

If P is not a ball-like point, we write

$$\alpha = e_1 = \sum a_i \partial_i,\ \beta = e_2 = \sum b_i \partial_i$$

$$S_{1\bar{1}1\bar{1}} = R(e_1, \bar{e}_1, e_1, \bar{e}_1) = \sum R_{i\bar{j}k\bar{l}} a_i \bar{a}_j a_k \bar{a}_l$$

and so on.

In particular, we have

$$S_{1\bar{1}1\bar{1}} = S_{2\bar{2}2\bar{2}},\ S_{1\bar{1}1\bar{2}} = S_{2\bar{2}2\bar{1}} = 0$$

We calculate the Laplace of $\Phi_1 = \tau^2 = \frac{|B|^2}{A^2}$ at a critical point.

We let

$$x_i = \nabla_i \Phi_1 = 2\frac{\tau}{A}[\mathrm{Re}\nabla_i S_{1\bar{2}1\bar{2}} + 3\tau \nabla_i S_{1\bar{1}1\bar{1}}]$$

As in [1,3,5], we have

$$\Delta R_{1\bar{1}1\bar{1}} = -AR_{1\bar{1}2\bar{2}} + B^2$$

$$\Delta R_{1\bar{2}1\bar{2}} = 3(R_{1\bar{1}2\bar{2}} - A)B$$

At P, we have $a_1 = b_2 = 1$ and $a_2 = b_1 = 0$, $\nabla a_1 = \nabla b_2 = 0$, $\nabla a_2 + \nabla \bar{b}_1 = 0$. Therefore, we write $y_{i1} = \nabla_i a_2$ and $y_{i2} = \nabla_i \bar{a}_2$. We also have

$$\Delta(a_1 + \bar{a}_1) = -|\nabla a_2|^2, \Delta(a_2 + \bar{b}_2) = 0$$

$$\nabla_i R_{1\bar{1}1\bar{2}} = -Ay_{i1} - By_{i2}$$

because

$$0 = \nabla S_{1\bar{1}1\bar{2}} = \nabla R_{1\bar{1}1\bar{2}} + 2R_{2\bar{1}1\bar{2}} \nabla a_2 + B\nabla \bar{a}_2 + R_{1\bar{1}1\bar{1}} \nabla \bar{b}_1$$

that is,

$$\nabla R_{1\bar{1}1\bar{2}} = -A\nabla a_2 - B\nabla \bar{a}_2$$

This also gives a similar formula for $\nabla_{\bar{i}} R_{1\bar{1}1\bar{2}}$. Similarly,

$$\nabla S_{1\bar{1}1\bar{1}} = \nabla R_{1\bar{1}1\bar{1}}$$

$$\nabla S_{1\bar{2}1\bar{2}} = \nabla R_{1\bar{2}1\bar{2}}$$

$$\Delta S_{1\bar{1}1\bar{1}} = -2A\sum |y|^2 - 4B\mathrm{Re}\sum y_{i1}\bar{y}_{i2} - AR_{1\bar{1}2\bar{2}} + B^2$$

$$\mathrm{Re}\Delta S_{1\bar{2}1\bar{2}} = 4A\sum \mathrm{Re} y_{i1}\bar{y}_{i2} + 2B\sum |y|^2 + 3(R_{1\bar{1}2\bar{2}} - A)B$$

$$\nabla_{\bar{1}} S_{1\bar{2}1\bar{2}} = -A\bar{y}_{22} - B\bar{y}_{21}$$

$$\nabla_2 S_{1\bar{2}1\bar{2}} = Ay_{11} + By_{12}$$

$$\nabla_1 S_{1\bar{2}1\bar{2}} = -A(6\tau^2 - 1)y_{22} - 5A\tau y_{21} + x_1$$

$$\nabla_{\bar{2}} S_{1\bar{2}1\bar{2}} = 5A\tau \bar{y}_{12} + A(6\tau^2 - 1)\bar{y}_{11} + \bar{x}_2$$

As in [3] (p. 598), at P we have

$$
\begin{aligned}
\Delta\Phi_1 &= \frac{2\tau\Delta B}{A} + \frac{6\tau^2}{A}\Delta S_{1\bar{1}1\bar{1}} \\
&+ \tfrac{1}{A^2}\sum(|\nabla S_{1\bar{2}1\bar{2}}|^2 + |\bar{\nabla} S_{1\bar{2}1\bar{2}}|^2) + \tfrac{54\tau^2}{A^2}\sum|\nabla S_{1\bar{1}\bar{1}}|^2 \\
&+ \tfrac{12\tau}{A^2}\sum \mathrm{Re}(\nabla_i S_{1\bar{1}1\bar{1}}(\nabla_{\bar{i}}(S_{1\bar{2}1\bar{2}} + S_{2\bar{1}2\bar{1}})) \\
&= 2\tau[3A\tau(\tau^2-1) - 4\tau\sum|y|^2 + 4(1-3\tau^2)\sum \mathrm{Re}(y_{i1}\bar{y}_{i2})] \\
&+ |y_{22} + \tau y_{21}|^2 + |y_{11} + \tau y_{12}|^2 \\
&+ \tfrac{1}{A^2}[|x_1 + A[(1-6\tau^2)y_{22} - 5\tau y_{21}]|^2 + |x_2 + A[(6\tau^2-1)y_{11} + 5\tau y_{12}]|^2 \\
&- 18\tau^2[y_{12} + \tau y_{11}|^2 + |y_{21} + \tau y_{22}|^2] \\
&+ \tfrac{12\tau}{A}[\mathrm{Re}[(y_{21} + \tau y_{22})\bar{x}_1] - \mathrm{Re}[(y_{21} + \tau y_{11})\bar{x}_2]]
\end{aligned} \tag{5}
$$

Here we notice that $\Delta\Phi_1$ has two general terms. The first term has nothing to do with x or y and therefore can be regarded as a constant term to these. This term is always nonpositive, because $\frac{1}{3} \leq \tau \leq 1$.

The second term can be regarded as a Hermitian form h to x and y. We can separate x and y into two groups: x_1, y_{2j} in one group and x_2, y_{1j} in the other. These two groups of variables are orthogonal to each other with respect to this Hermitian form. That is, $h = h_1 + h_2$ with h_1 (or h_2) only depends on the first (second) group of variables.

We need to check the nonpositivity for each of these.

For x_2, y_{11}, y_{12}, the corresponding matrix of h_2 is

$$
\begin{bmatrix}
\frac{1}{A^2} & -\frac{1}{A} & -\frac{\tau}{A} \\
-\frac{1}{A} & 2(9\tau^2-1)(\tau^2-1) & 0 \\
-\frac{\tau}{A} & 0 & 0
\end{bmatrix}
$$

The matrix for h_1 of x_1, y_{21}, y_{22} is

$$
\begin{bmatrix}
\frac{1}{A^2} & \frac{\tau}{A} & \frac{1}{A} \\
\frac{\tau}{A} & 0 & 0 \\
\frac{1}{A} & 0 & 2(9\tau^2-1)(\tau^2-1)
\end{bmatrix}
$$

When P is a critical point of Φ_1, then $x_1 = x_2 = 0$. The matrices on y are clearly semidefinite.

Acknowledgments: I thank Professors Poon and Wong and the Department of Mathematics, University of California at Riverside for their support. I thank Professor Hong-Cang Yang for showing me his work when I was a graduate student in Berkeley. I also thank Professor Paul Yang for sharing with me the work [4]. I also thank Tommy Murphy for discussions and the referees of [6] and this paper for their useful comments. Finally, this paper was written up when I visited Xiamen University. Here I take this chance to thank Professor Qiu Chunhui and the department for their warm hospitality.

Conflicts of Interest: The author declares no conflict of interest.

References

1. Siu, Y.T.; Yang, P. Compact Kähler-Einstein surfaces of nonpositive bisectional curvature. *Invent. Math.* **1981**, *64*, 471–487.
2. Polombo, A. De Nouvelles Formules De Weitzenböck Pour Des Endomorphismes Harmoniques Applications Géométriques. *Ann. Sci. Ec. Norm.* **1992**, *25*, 393–428.
3. Hong, Y.; Guan, Z.D.; Yang, H.C. A note on the Kähler-Einstein Manifolds. *Acta. Math. Sin.* **1988**, *31*, 595–602.
4. Polombo, A. Condition d'Einstein et courbure negative en dimension 4. *C. R. Acad. Sci. Paris Ser. 1 Math.* **1988**, *307*, 667–670.

5. Chen, D.; Hong, Y.; Yang, H.C. Kähler-Einstein surface and symmetric space. *Sci. China Math.* **2011**, *54*, 2627–2634.
6. Guan, D. On Bisectional Negatively Curved Compact Kähler-Einstein Surfaces. *Pac. J. Math.* **2017**, *288*, 343–353.
7. Dancer, A.; Wang, M. Kähler Einstein Metrics of Cohomogeneity One and Bundle Construction for Einstein Hermitian Metrics. *Math. Ann.* **1998**, *312*, 503–526.
8. Apostolov, V.; Calderbank, D.; Gauduchon, P.C. Tonnesen-Friedman: Hamilton 2-forms in Kähler geometry III, extremal metrics and stability. *Invent. Math.* **2008**, *173*, 547–601.
9. Bogomolov, F. On Guan's examples of simply connected non-kähler compact complex manifolds. *Am. J. Math.* **1996**, *118*, 1037–1046.
10. Console, S.; Fino, A. On the de Rham cohomology of solvmanifolds. *Ann. Scuola Norm. Super. Pisa* **2011**, *10*, 801–818.
11. Guan, D. A Note On the classification of Compact Complex Homogeneous Locally Conformal Kähler Manifolds. *J. Math. Stat.* **2017**, *13*, 261–267.
12. Huckleberry, A. Homogeneous Pseudo-Kähler Manifolds: A Hamiltonian Viewpoint. *Note Math.* **1990**, *10* (Suppl. 2), 337–342.
13. Podesta, F.; Spiro, A. Kähler manifolds with large isometry group. *Osaka J. Math.* **1999**, *36*, 805–833.
14. Tian, G. Smoothness of the Universal Deformations Space of Compact Calabi-Yau Manifolds and Its Peterson-Well Metric. In *Mathematical Aspect of String Theory*; Yau, S.T., ed.; World Scientific: Singapore, 1987; pp. 629–646.
15. Mok, N.; Zhong, J.Q. Curvature Characterization of Compact Hermitian Symmetric Spaces. *J. Diff. Geom.* **1986**, *23*, 15–67.
16. Derdzinski, A. Self-dual Kähler manifolds and Einstein manifolds of dimension four. *Compos. Math.* **1983**, *49*, 405–433.

mathematics

MDPI

Article

Comparison of Differential Operators with Lie Derivative of Three-Dimensional Real Hypersurfaces in Non-Flat Complex Space Forms

George Kaimakamis [1], Konstantina Panagiotidou [1,*] and Juan de Dios Pérez [2]

[1] Faculty of Mathematics and Engineering Sciences, Hellenic Army Academy, Varia, 16673 Attiki, Greece; gmiamis@gmail.com or gmiamis@sse.gr
[2] Departamento de Geometria y Topologia, Universidad de Granada, 18071 Granada, Spain; jdperez@ugr.es
* Correspondence: konpanagiotidou@gmail.com
Received: 29 March 2018; Accepted: 14 May 2018; Published: 20 May 2018

Abstract: In this paper, three-dimensional real hypersurfaces in non-flat complex space forms, whose shape operator satisfies a geometric condition, are studied. Moreover, the tensor field $P = \phi A - A\phi$ is given and three-dimensional real hypersurfaces in non-flat complex space forms whose tensor field P satisfies geometric conditions are classified.

Keywords: k-th generalized Tanaka–Webster connection; non-flat complex space form; real hypersurface; lie derivative; shape operator

2010 Mathematics Subject Classification: 53C15; 53B25

1. Introduction

A *real hypersurface* is a submanifold of a Riemannian manifold with a real co-dimensional one. Among the Riemannian manifolds, it is of great interest in the area of Differential Geometry to study real hypersurfaces in complex space forms. A *complex space form* is a Kähler manifold of dimension n and constant holomorphic sectional curvature c. In addition, complete and simply connected complex space forms are analytically isometric to complex projective space $\mathbb{C}P^n$ if $c > 0$, to complex Euclidean space $\mathbb{C}^n$ if $c = 0$, or to complex hyperbolic space $\mathbb{C}H^n$ if $c < 0$. The notion of non-flat complex space form refers to complex projective and complex hyperbolic space when it is not necessary to distinguish between them and is denoted by $M_n(c), n \geq 2$.

Let J be the Kähler structure and $\tilde{\nabla}$ the Levi–Civita connection of the non-flat complex space form $M_n(c), n \geq 2$. Consider M a connected real hypersurface of $M_n(c)$ and N a locally defined unit normal vector field on M. The Kähler structure induces on M an *almost contact metric structure* (ϕ, ξ, η, g). The latter consists of a tensor field of type (1, 1) ϕ called *structure tensor field*, a one-form η, a vector field ξ given by $\xi = -JN$ known as the *structure vector field* of M and g, which is the induced Riemannian metric on M by G. Among real hypersurfaces in non-flat complex space forms, the class of *Hopf hypersurfaces* is the most important. A Hopf hypersurface is a real hypersurface whose structure vector field ξ is an eigenvector of the shape operator A of M .

Takagi initiated the study of real hypersurfaces in non-flat complex space forms. He provided the classification of homogeneous real hypersurfaces in complex projective space $\mathbb{C}P^n$ and divided them into five classes (A), (B), (C), (D) and (E) (see [1–3]). Later, Kimura proved that homogeneous real hypersurfaces in complex projective space are the unique Hopf hypersurfaces with constant principal curvatures, i.e., the eigenvalues of the shape operator A are constant (see [4]). Among the above real hypersurfaces, the three-dimensional real hypersurfaces in $\mathbb{C}P^2$ are geodesic hyperspheres of radius r, $0 < r < \frac{\pi}{2}$, called real hypersurfaces of type (A) and tubes of radius r, $0 < r < \frac{\pi}{4}$, over the complex quadric called real hypersurfaces of type (B). Table 1 includes the values of the constant principal curvatures corresponding to the real hypersurfaces above (see [1,2]).

Table 1. Principal curvatures of real hypersurfaces in $\mathbb{C}P^2$.

Type	α	λ_1	ν	m_α	m_{λ_1}	m_ν
(A)	$2\cot(2r)$	$\cot(r)$	-	1	2	-
(B)	$2\cot(2r)$	$\cot(r-\frac{\pi}{4})$	$-\tan(r-\frac{\pi}{4})$	1	1	1

The study of Hopf hypersurfaces with constant principal curvatures in complex hyperbolic space $\mathbb{C}H^n, n \geq 2$, was initiated by Montiel in [5] and completed by Berndt in [6]. They are divided into two types: type (A), which are open subsets of horospheres (A_0), geodesic hyperspheres $(A_{1,0})$, or tubes over totally geodesic complex hyperbolic hyperplane $\mathbb{C}H^{n-1}$ $(A_{1,1})$ and type (B), which are open subsets of tubes over totally geodesic real hyperbolic space $\mathbb{R}H^n$. Table 2 includes the values of the constant principal curvatures corresponding to above real hypersurfaces for $n = 2$ (see [6]).

Table 2. Principal curvatures of real hypersurfaces in $\mathbb{C}H^2$.

Type	α	λ	ν	m_α	m_λ	m_ν
(A_0)	2	1	-	1	2	-
$(A_{1,1})$	$2\coth(2r)$	$\coth(r)$	-	1	2	-
$(A_{1,2})$	$2\coth(2r)$	$\tanh(r)$	-	1	2	-
(B)	$2\tanh(2r)$	$\tanh(r)$	$\coth(r)$	1	1	1

The Levi–Civita connection $\tilde{\nabla}$ of the non-flat complex space form $M_n(c), n \geq 2$ induces on M a Levi–Civita connection ∇. Apart from the last one, Cho in [7,8] introduces the notion of the *k-th generalized Tanaka–Webster connection* $\hat{\nabla}^{(k)}$ on a real hypersurface in non-flat complex space form given by

$$\hat{\nabla}^{(k)}_X Y = \nabla_X Y + g(\phi AX, Y)\xi - \eta(Y)\phi AX - k\eta(X)\phi Y, \tag{1}$$

for all X, Y tangent to M , where k is a nonnull real number. The latter is an extension of the definition of *generalized Tanaka–Webster connection* for contact metric manifolds given by Tanno in [9] and satisfying the relation

$$\hat{\nabla}_X Y = \nabla_X Y + (\nabla_X \eta)(Y)\xi - \eta(Y)\nabla_X \xi - \eta(X)\phi Y.$$

The following relations hold:

$$\hat{\nabla}^{(k)}\eta = 0, \quad \hat{\nabla}^{(k)}\xi = 0, \quad \hat{\nabla}^{(k)}g = 0, \quad \hat{\nabla}^{(k)}\phi = 0.$$

In particular, if the shape operator of a real hypersurface satisfies $\phi A + A\phi = 2k\phi$, the generalized Tanaka–Webster connection coincides with the Tanaka–Webster connection.

The k-th Cho operator on M associated with the vector field X is denoted by $\hat{F}^{(k)}_X$ and given by

$$\hat{F}^{(k)}_X Y = g(\phi AX, Y)\xi - \eta(Y)\phi AX - k\eta(X)\phi Y, \tag{2}$$

for any Y tangent to M. Then, the torsion of the k-th generalized Tanaka–Webster connection $\hat{\nabla}^{(k)}$ is given by

$$T^{(k)}(X,Y) = \hat{F}^{(k)}_X Y - \hat{F}^{(k)}_Y X,$$

for any X, Y tangent to M. Associated with the vector field X, the *k-th torsion operator* $T^{(k)}_X$ is defined and given by

$$T_X^{(k)}Y = T^{(k)}(X,Y),$$

for any Y tangent to M.

The existence of Levi–Civita and k-th generalized Tanaka–Webster connections on a real hypersurface implies that the covariant derivative can be expressed with respect to both connections. Let K be a tensor field of type (1, 1); then, the symbols ∇K and $\hat{\nabla}^{(k)}K$ are used to denote the covariant derivatives of K with respect to the Levi–Civita and the k-th generalized Tanaka–Webster connection, respectively. Furthermore, the Lie derivative of a tensor field K of type (1, 1) with respect to Levi–Civita connection $\mathcal{L}K$ is given by

$$(\mathcal{L}_X K)Y = \nabla_X(KY) - \nabla_{KY}X - K\nabla_X Y + K\nabla_Y X, \tag{3}$$

for all X, Y tangent to M . Another first order differential operator of a tensor field K of type (1, 1) with respect to the k-th generalized Tanaka–Webster connection $\hat{\mathcal{L}}^{(k)}K$ is defined and it is given by

$$(\hat{\mathcal{L}}_X^{(k)}K)Y = \hat{\nabla}_X^{(k)}(KY) - \hat{\nabla}_{KY}^{(k)}X - K(\hat{\nabla}_X^{(k)}Y) + K(\hat{\nabla}_Y^{(k)}X), \tag{4}$$

for all X, Y tangent to M .

Due to the existence of the above differential operators and derivatives, the following questions come up

1. Are there real hypersurfaces in non-flat complex space forms whose derivatives with respect to different connections coincide?
2. Are there real hypersurfaces in non-flat complex space forms whose differential operator $\hat{\mathcal{L}}^{(k)}$ coincides with derivatives with respect to different connections?

The first answer is obtained in [10], where the classification of real hypersurfaces in complex projective space $\mathbb{C}P^n$,$n \geq 3$, whose covariant derivative of the shape operator with respect to the Levi–Civita connection coincides with the covariant derivative of it with respect to the k-th generalized Tanaka–Webster connection is provided, i.e., $\nabla_X A = \hat{\nabla}_X^{(k)}A$, where X is any vector field on M. Next, in [11], real hypersurfaces in complex projective space $\mathbb{C}P^n$, $n \geq 3$, whose Lie derivative of the shape operator coincides with the operator $\hat{\mathcal{L}}^{(k)}$ are studied, i.e., $\mathcal{L}_X A = \hat{\mathcal{L}}_X^{(k)}A$, where X is any vector field on M. Finally, in [12], the problem of classifying three-dimensional real hypersurfaces in non-flat complex space forms $M_2(c)$, for which the operator $\hat{\mathcal{L}}^{(k)}$ applied to the shape operator coincides with the covariant derivative of it, has been studied, i.e., $\hat{\mathcal{L}}_X^{(k)}A = \nabla_X A$, for any vector field X tangent to M.

In this paper, the condition $\mathcal{L}_X A = \hat{\mathcal{L}}_X^{(k)}A$, where X is any vector field on M is studied in the case of three-dimensional real hypersurfaces in $M_2(c)$.

The aim of the present paper is to complete the work of [11] in the case of three-dimensional real hypersurfaces in non-flat complex space forms $M_2(c)$. The equality $\mathcal{L}_X A = \hat{\mathcal{L}}_X^{(k)}A$ is equivalent to the fact that $T_X^{(k)}A = AT_X^{(k)}$. Thus, the eigenspaces of A are preserved by the k-th torsion operator $T_X^{(k)}$, for any X tangent to M . First, three-dimensional real hypersurfaces in $M_2(c)$ whose shape operator A satisfies the following relation:

$$\hat{\mathcal{L}}_X^{(k)}A = \mathcal{L}_X A, \tag{5}$$

for any X orthogonal to ξ are studied and the following Theorem is proved:

Theorem 1. *There do not exist real hypersurfaces in $M_2(c)$ whose shape operator satisfies relation (5).*

Next, three-dimensional real hypersurfaces in $M_2(c)$ whose shape operator satisfies the following relation are studied:

$$\hat{\mathcal{L}}_{\xi}^{(k)} A = \mathcal{L}_{\xi} A, \tag{6}$$

and the following Theorem is provided.

Theorem 2. *Every real hypersurface in $M_2(c)$ whose shape operator satisfies relation (6) is locally congruent to a real hypersurface of type (A).*

As an immediate consequence of the above theorems, it is obtained that

Corollary 1. *There do not exist real hypersurfaces in $M_2(c)$ such that $\hat{\mathcal{L}}_X^{(k)} A = \mathcal{L}_X A$, for all $X \in TM$.*

Next, the following tensor field P of type $(1,1)$ is introduced:

$$PX = \phi AX - A\phi X,$$

for any vector field X tangent to M. The relation $P = 0$ implies that the shape operator commutes with the structure tensor ϕ. Real hypersurfaces whose shape operator A commutes with the structure tensor ϕ have been studied by Okumura in the case of $\mathbb{C}P^n$, $n \geq 2$, (see [13]) and by Montiel and Romero in the case of $\mathbb{C}H^n$, $n \geq 2$ (see [14]). The following Theorem provides the above classification of real hypersurfaces in $M_n(c)$, $n \geq 2$.

Theorem 3. *Let M be a real hypersurface of $M_n(c)$, $n \geq 2$. Then, $A\phi = \phi A$, if and only if M is locally congruent to a homogeneous real hypersurface of type (A). More precisely:*
In the case of $\mathbb{C}P^n$

(A_1) *a geodesic hypersphere of radius r, where $0 < r < \frac{\pi}{2}$,*
(A_2) *a tube of radius r over a totally geodesic $\mathbb{C}P^k$, $(1 \leq k \leq n-2)$, where $0 < r < \frac{\pi}{2}$.*

In the case of $\mathbb{C}H^n$,

(A_0) *a horosphere in $\mathbb{C}H^n$, i.e., a Montiel tube,*
(A_1) *a geodesic hypersphere or a tube over a totally geodesic complex hyperbolic hyperplane $\mathbb{C}H^{n-1}$,*
(A_2) *a tube over a totally geodesic $\mathbb{C}H^k$ $(1 \leq k \leq n-2)$.*

Remark 1. *In the case of three-dimensional real hypersurfaces in $M_2(c)$, real hypersurfaces of type (A_2) do not exist.*

It is interesting to study real hypersurfaces in non-flat complex spaces forms, whose tensor field P satisfies certain geometric conditions. We begin by studying three-dimensional real hypersurfaces in $M_2(c)$ whose tensor field P satisfies the relation

$$(\hat{\mathcal{L}}_X^{(k)} P)Y = (\mathcal{L}_X P)Y, \tag{7}$$

for any vector fields X, Y tangent to M.

First, the following Theorem is proved:

Theorem 4. *Every real hypersurface in $M_2(c)$ whose tensor field P satisfies relation (7) for any X orthogonal to ξ and $Y \in TM$ is locally congruent to a real hypersurface of type (A).*

Next, we study three-dimensional real hypersurfaces in $M_2(c)$ whose tensor field P satisfies relation (7) for $X = \xi$, i.e.,

$$(\hat{\mathcal{L}_{\xi}^{(k)}} P)Y = (\mathcal{L}_{\xi} P)Y, \tag{8}$$

for any vector field Y tangent to M. Then, the following Theorem is proved:

Theorem 5. *Every real hypersurface in $M_2(c)$ whose tensor field P satisfies relation (8) is a Hopf hypersurface. In the case of $\mathbb{C}P^2$, M is locally congruent to a real hypersurface of type (A) or to a real hypersurface of type (B) with $\alpha = -2k$ and in the case of $\mathbb{C}H^2$ M is a locally congruent either to a real hypersurface of type (A) or to a real hypersurface of type (B) with $\alpha = \frac{4}{k}$.*

This paper is organized as follows: in Section 2, basic relations and theorems concerning real hypersurfaces in non-flat complex space forms are presented. In Section 3, analytic proofs of Theorems 1 and 2 are provided. Finally, in Section 4, proofs of Theorems 4 and 5 are given.

2. Preliminaries

Throughout this paper, all manifolds, vector fields, etc. are considered of class C^∞ and all manifolds are assumed to be connected.

The non-flat complex space form $M_n(c)$, $n \geq 2$ is equipped with a Kähler structure J and G is the Kählerian metric. The constant holomorphic sectional curvature c in the case of complex projective space $\mathbb{C}P^n$ is $c = 4$ and in the case of complex hyperbolic space $\mathbb{C}H^n$ is $c = -4$. The Levi–Civita connection of the non-flat complex space form is denoted by $\overline{\nabla}$.

Let M be a connected real hypersurface immersed in $M_n(c)$, $n \geq 2$, without boundary and N be a locally defined unit normal vector field on M. The shape operator A of the real hypersurface M with respect to the vector field N is given by

$$\overline{\nabla}_X N = -AX.$$

The Levi–Civita connection ∇ of the real hypersurface M satisfies the relation

$$\overline{\nabla}_X Y = \nabla_X Y + g(AX, Y)N.$$

The Kähler structure of the ambient space induces on M an almost contact metric structure (ϕ, ξ, η, g) in the following way: any vector field X tangent to M satisfies the relation

$$JX = \phi X + \eta(X)N.$$

The tangential component of the above relation defines on M a skew-symmetric tensor field of type $(1,1)$ denoted by ϕ known as *the structure tensor*. The structure vector field ξ is defined by $\xi = -JN$ and the 1-form η is given by $\eta(X) = g(X, \xi)$ for any vector field X tangent to M. The elements of the almost contact structure satisfy the following relation:

$$\phi^2 X = -X + \eta(X)\xi, \quad \eta(\xi) = 1, \quad g(\phi X, \phi Y) = g(X, Y) - \eta(X)\eta(Y) \tag{9}$$

for all tangent vectors X, Y to M. Relation (9) implies

$$\phi\xi = 0, \quad \eta(X) = g(X, \xi).$$

Because of $\overline{\nabla} J = 0$, it is obtained

$$(\nabla_X \phi)Y = \eta(Y)AX - g(AX, Y)\xi \quad \text{and} \quad \nabla_X \xi = \phi AX$$

for all X, Y tangent to M. Moreover, the Gauss and Codazzi equations of the real hypersurface are respectively given by

$$R(X,Y)Z = \frac{c}{4}[g(Y,Z)X - g(X,Z)Y + g(\phi Y,Z)\phi X - g(\phi X,Z)\phi Y \\ -2g(\phi X,Y)\phi Z] + g(AY,Z)AX - g(AX,Z)AY, \tag{10}$$

and

$$(\nabla_X A)Y - (\nabla_Y A)X = \frac{c}{4}[\eta(X)\phi Y - \eta(Y)\phi X - 2g(\phi X,Y)\xi], \tag{11}$$

for all vectors X, Y, Z tangent to M, where R is the curvature tensor of M.

The tangent space T_pM at every point $p \in M$ is decomposed as

$$T_pM = span\{\xi\} \oplus \mathbb{D}, \tag{12}$$

where $\mathbb{D} = \ker\eta = \{X \in T_pM : \eta(X) = 0\}$ and is called *(maximal) holomorphic distribution (if $n \geq 3$)*.

Next, the following results concern any non-Hopf real hypersurface M in $M_2(c)$ with local orthonormal basis $\{U, \phi U, \xi\}$ at a point p of M.

Lemma 1. *Let M be a non-Hopf real hypersurface in $M_2(c)$. The following relations hold on M:*

$$\begin{array}{lll} AU = \gamma U + \delta\phi U + \beta\xi, & A\phi U = \delta U + \mu\phi U, & A\xi = \alpha\xi + \beta U, \\ \nabla_U\xi = -\delta U + \gamma\phi U, & \nabla_{\phi U}\xi = -\mu U + \delta\phi U, & \nabla_\xi\xi = \beta\phi U, \\ \nabla_U U = \kappa_1\phi U + \delta\xi, & \nabla_{\phi U}U = \kappa_2\phi U + \mu\xi, & \nabla_\xi U = \kappa_3\phi U, \\ \nabla_U\phi U = -\kappa_1 U - \gamma\xi, & \nabla_{\phi U}\phi U = -\kappa_2 U - \delta\xi, & \nabla_\xi\phi U = -\kappa_3 U - \beta\xi, \end{array} \tag{13}$$

where $\alpha, \beta, \gamma, \delta, \mu, \kappa_1, \kappa_2, \kappa_3$ are smooth functions on M and $\beta \neq 0$.

Remark 2. *The proof of Lemma 1 is included in [15].*

The Codazzi equation for $X \in \{U, \phi U\}$ and $Y = \xi$ implies, because of Lemma 1, the following relations:

$$\xi\delta = \alpha\gamma + \beta\kappa_1 + \delta^2 + \mu\kappa_3 + \frac{c}{4} - \gamma\mu - \gamma\kappa_3 - \beta^2, \tag{14}$$

$$\xi\mu = \alpha\delta + \beta\kappa_2 - 2\delta\kappa_3, \tag{15}$$

$$(\phi U)\alpha = \alpha\beta + \beta\kappa_3 - 3\beta\mu, \tag{16}$$

$$(\phi U)\beta = \alpha\gamma + \beta\kappa_1 + 2\delta^2 + \frac{c}{2} - 2\gamma\mu + \alpha\mu, \tag{17}$$

and for $X = U$ and $Y = \phi U$

$$U\delta - (\phi U)\gamma = \mu\kappa_1 - \kappa_1\gamma - \beta\gamma - 2\delta\kappa_2 - 2\beta\mu. \tag{18}$$

The following Theorem refers to Hopf hypersurfaces. In the case of complex projective space $\mathbb{C}P^n$, it is given by Maeda [16], and, in the case of complex hyperbolic space $\mathbb{C}H^n$, it is given by Ki and Suh [17] (see also Corollary 2.3 in [18]).

Theorem 6. *Let M be a Hopf hypersurface in $M_n(c)$, $n \geq 2$. Then,*

(i) $\alpha = g(A\xi, \xi)$ *is constant.*

(ii) *If W is a vector field, which belongs to $\mathbb{D}$ such that $AW = \lambda W$, then*

$$(\lambda - \frac{\alpha}{2})A\phi W = (\frac{\lambda\alpha}{2} + \frac{c}{4})\phi W.$$

(iii) *If the vector field W satisfies $AW = \lambda W$ and $A\phi W = \nu\phi W$, then*

$$\lambda\nu = \frac{\alpha}{2}(\lambda + \nu) + \frac{c}{4}. \tag{19}$$

Remark 3. *Let M be a three-dimensional Hopf hypersurface in $M_2(c)$. Since M is a Hopf hypersurface relation $A\xi = \alpha\xi$, it holds when $\alpha = constant$. At any point $p \in M$, we consider a unit vector field $W \in \mathbb{D}$ such that $AW = \lambda W$. Then, the unit vector field ϕW is orthogonal to W and ξ and relation $A\phi W = \nu\phi W$ holds. Therefore, at any point $p \in M$, we can consider the local orthonormal frame $\{W, \phi W, \xi\}$ and the shape operator satisfies the above relations.*

3. Proofs of Theorems 1 and 2

Suppose that M is a real hypersurface in $M_2(c)$ whose shape operator satisfies relation (5), which because of the relation of k-th generalized Tanaka-Webster connection (1) becomes

$$\begin{aligned} & g((A\phi A + A^2\phi)X, Y)\xi - g((A\phi + \phi A)X, Y)A\xi + k\eta(AY)\phi X + \eta(Y)A\phi AX \\ & -\eta(AY)\phi AX - k\eta(Y)A\phi X = 0, \end{aligned} \tag{20}$$

for any $X \in \mathbb{D}$ and for all $Y \in TM$.

Let $\mathcal{N}$ be the open subset of M such that

$$\mathcal{N} = \{p \in M : \beta \neq 0, \text{ in a neighborhood of } p\}.$$

The inner product of relation (20) for $Y = \xi$ with ξ due to relation (13) implies $\delta = 0$ and the shape operator on the local orthonormal basis $\{U, \phi U, \xi\}$ becomes

$$A\xi = \alpha\xi + \beta U, \quad AU = \gamma U + \beta\xi \text{ and } A\phi U = \mu\phi U. \tag{21}$$

Relation (20) for $X = Y = U$ and $X = \phi U$ and $Y = \xi$ due to (21) yields, respectively,

$$\gamma = k \text{ and } \mu = 0. \tag{22}$$

Differentiation of $\gamma = k$ with respect to ϕU taking into account that k is a nonzero real number implies $(\phi U)\gamma = 0$. Thus, relation (18) results, because of $\delta = \mu = 0$, in $\kappa_1 = -\beta$. Furthermore, relations (14)–(17) due to $\delta = 0$ and relation (22) become

$$\alpha k + \frac{c}{4} = 2\beta^2 + k\kappa_3, \tag{23}$$

$$\kappa_2 = 0, \tag{24}$$

$$(\phi U)\alpha = \beta(\alpha + \kappa_3), \tag{25}$$

$$(\phi U)\beta = \alpha k - \beta^2 + \frac{c}{2}. \tag{26}$$

The inner product of Codazzi equation (11) for $X = U$ and $Y = \xi$ with U and ξ implies because of $\delta = 0$ and relation (21),

$$U\alpha = U\beta = \xi\beta = \xi\gamma = 0. \tag{27}$$

The Lie bracket of U and ξ satisfies the following two relations:

$$[U, \xi]\beta = U(\xi\beta) - \xi(U\beta),$$
$$[U, \xi]\beta = (\nabla_U\xi - \nabla_\xi U)\beta.$$

A combination of the two relations above taking into account relations of Lemma 1 and (27) yields

$$(k-\kappa_3)[(\phi U)\beta]=0.$$

Suppose that $k\neq\kappa_3$, then $(\phi U)\beta=0$ and relation (26) implies $\alpha k+\frac{c}{2}=\beta^2$. Differentiation of the last one with respect to ϕU results, taking into account relation (25), in $\kappa_3=-\alpha$. The Riemannian curvature satisfies the relation

$$R(X,Y)Z=\nabla_X\nabla_Y Z-\nabla_Y\nabla_X Z-\nabla_{[X,Y]}Z,$$

for any X, Y, Z tangent to M. Combination of the last relation with Gaussian Equation (10) for $X=U$, $Y=\phi U$ and $Z=U$ due to relation (22) and relation (24), $\kappa_1=-\beta$, $\kappa_3=-\alpha$ and $(\phi U)\beta=0$ implies $c=0$, which is a contradiction.

Therefore, on M, relation $k=\kappa_3$ holds. A combination of $R(X,Y)Z=\nabla_X\nabla_Y Z-\nabla_Y\nabla_X Z-\nabla_{[X,Y]}Z$ with Gauss Equation (10) for $X=U$, $Y=\phi U$ and $Z=U$ because of relations (22) and (26) and $\kappa_1=-\beta$ yields

$$k^2=-\alpha k-\frac{3c}{2}.$$

A combination of the latter with relation (23) implies

$$\beta^2+k^2=-\frac{5c}{8}.$$

Differentiation of the above relation with respect to ϕU gives, due to relation (26) and $k^2=-\alpha k-\frac{3c}{2}$,

$$\beta^2+k^2=-\frac{c}{2}.$$

If the ambient space is the complex projective space $\mathbb{C}P^2$ with $c=4$, then the above relation leads to a contradiction. If the ambient space is the complex hyperbolic space $\mathbb{C}H^2$ with $c=-4$, combination of the latter relation with $\beta^2+k^2=-\frac{5c}{8}$ yields $c=0$, which is a contradiction.

Thus, $\mathcal{N}$ is empty and the following proposition is proved:

Proposition 1. *Every real hypersurface in $M_2(c)$ whose shape operator satisfies relation (5) is a Hopf hypersurface.*

Since M is a Hopf hypersurface, Theorem 6 and remark 3 hold. Relation (20) for $X=W$ and for $X=\phi W$ implies, respectively,

$$(\lambda-k)(\nu-\alpha)=0 \text{ and } (\nu-k)(\lambda-\alpha)=0. \tag{28}$$

Combination of the above relations results in

$$(\nu-\lambda)(\alpha-k)=0.$$

If $\lambda\neq\nu$, then $\alpha=k$ and relation $(\lambda-k)(\nu-\alpha)=0$ becomes

$$(\lambda-\alpha)(\nu-\alpha)=0.$$

If $\nu\neq\alpha$, then $\lambda=\alpha$ and relation (19) implies that ν is also constant. Therefore, the real hypersurface is locally congruent to a real hypersurface of type (B). Substitution of the values of eigenvalues in relation $\lambda = \alpha$ leads to a contradiction. Thus, on M, relation $\nu = \alpha$ holds. Following similar steps to the previous case, we are led to a contradiction.

Therefore, on M, we have $\lambda = \nu$ and the first of relations (28) becomes

$$(\lambda - k)(\lambda - \alpha) = 0.$$

Supposing that $\lambda \neq k$, then $\lambda = \nu = \alpha$. Thus, the real hypersurface is totally umbilical, which is impossible since there do not exist totally umbilical real hypersurfaces in non-flat complex space forms [18].

Thus, on M relation $\lambda = k$ holds. Relation (20) for $X = W$ and $Y = \phi W$ implies, because of $\lambda = \nu = k$, $\lambda = \alpha$. Thus, $\lambda = \nu = \alpha$ and the real hypersurface is totally umbilical, which is a contradiction and this completes the proof of Theorem 1.

Next, suppose that M is a real hypersurface in $M_2(c)$ whose shape operator satisfies relation (6), which, because of the relation of the k-th generalized Tanaka-Webster connection (1), becomes

$$(A\phi - \phi A)AX - g(\phi A\xi, AX)\xi + \eta(AX)\phi A\xi + k\phi AX + g(\phi A\xi, X)A\xi$$
$$-\eta(X)A\phi A\xi - kA\phi X = 0, \tag{29}$$

for any $X \in TM$.

Let $\mathcal{N}$ be the open subset of M such that

$$\mathcal{N} = \{p \in M : \beta \neq 0, \text{ in a neighborhood of } p\}.$$

The inner product of relation (29) for $X = U$ with ξ implies, due to relation (13), $\delta = 0$ and the shape operator on the local orthonormal basis $\{U, \phi U, \xi\}$ becomes

$$A\xi = \alpha\xi + \beta U, \quad AU = \gamma U + \beta\xi \text{ and } A\phi U = \mu\phi U. \tag{30}$$

Relation (29) for $X = \xi$ yields, taking into account relation (30), $\gamma = k$. Finally, relation (29) for $X = \phi U$ implies, due to relation (30) and the last relation,

$$(\mu^2 - 2k\mu + k^2) + \beta^2 = 0.$$

The above relation results in $\beta = 0$, which implies that $\mathcal{N}$ is empty. Thus, the following proposition is proved:

Proposition 2. *Every real hypersurface in $M_2(c)$ whose shape operator satisfies relation (6) is a Hopf hypersurface.*

Due to the above Proposition, Theorem 6 and Remark 3 hold. Relation (29) for $X = W$ and for $X = \phi W$ implies, respectively,

$$(\lambda - k)(\lambda - \nu) = 0 \text{ and } (\nu - k)(\lambda - \nu) = 0.$$

Suppose that $\lambda \neq \nu$. Then, the above relations imply $\lambda = \nu = k$, which is a contradiction.

Thus, on M, relation $\lambda = \nu$ holds and this results in the structure tensor ϕ commuting with the shape operator A, i.e., $A\phi = \phi A$ and, because of Theorem 3 M , is locally congruent to a real hypersurface of type (A), and this completes the proof of Theorem 2.

4. Proof of Theorems 4 and 5

Suppose that M is a real hypersurface in $M_2(c)$ whose tensor field P satisfies relation (7) for any $X \in \mathbb{D}$ and for all $Y \in TM$. Then, the latter relation becomes, because of the relation of the k-th generalized Tanaka-Webster connection (1) and relations (3) and (4),

$$g(\phi AX, PY)\xi - \eta(PY)\phi AX - g(\phi APY, X)\xi + k\eta(PY)\phi X - g(\phi AX, Y)P\xi$$
$$+\eta(Y)P\phi AX + g(\phi AY, X)P\xi - k\eta(Y)P\phi X = 0, \tag{31}$$

for any $X \in \mathbb{D}$ and for all $Y \in TM$.

Let $\mathcal{N}$ be the open subset of M such that

$$\mathcal{N} = \{p \in M : \beta \neq 0, \text{ in a neighborhood of } p\}.$$

Relation (31) for $Y = \xi$ implies, taking into account relation (13),

$$\beta\{g(AX, U) + g(A\phi U, \phi X)\}\xi + P\phi AX + \beta^2 g(\phi U, X)\phi U - kP\phi X = 0, \tag{32}$$

for any $X \in \mathbb{D}$.

The inner product of relation (32) for $X = \phi U$ with ξ due to relation (13) yields $\delta = 0$. Moreover, the inner product of relation (32) for $X = \phi U$ with ϕU, taking into account relation (13) and $\delta = 0$, results in

$$\beta^2 + k(\gamma - \mu) = \mu(\gamma - \mu). \tag{33}$$

The inner product of relation (32) for $X = U$ with U gives, because of relation (13) and $\delta = 0$,

$$(\gamma - k)(\gamma - \mu) = 0.$$

Suppose that $\gamma \neq k$, then the above relation implies $\gamma = \mu$ and relation (33) implies $\beta = 0$, which is impossible.

Thus, relation $\gamma = k$ holds and relation (33) results in

$$\beta^2 + (\gamma - \mu)^2 = 0.$$

The latter implies $\beta = 0$, which is impossible.

Thus, $\mathcal{N}$ is empty and the following proposition has been proved:

Proposition 3. *Every real hypersurface in $M_2(c)$ whose tensor field P satisfies relation (7) is a Hopf hypersurface.*

As a result of the proposition above, Theorem 6 and remark 3 hold. Thus, relation (31) for $X = W$ and $Y = \xi$ and for $X = \phi W$ and $Y = \xi$ yields, respectively,

$$(\lambda - k)(\lambda - \nu) = 0 \text{ and } (\nu - k)(\lambda - \nu) = 0.$$

Supposing that $\lambda \neq \nu$, the above relations imply $\lambda = \nu = k$, which is a contradiction.

Therefore, relation $\lambda = \nu$ holds and this implies that $A\phi = \phi A$. Thus, because of Theorem 3, M is locally congruent to a real hypersurface of type (A) and this completes the proof of Theorem 4.

Next, we study three-dimensional real hypersurfaces in $M_2(c)$ whose tensor field P satisfies relation (8). The last relation becomes, due to relation (2),

$$F_\xi^{(k)} PY - PF_\xi^{(k)} Y + \phi APY - P\phi AY = 0, \tag{34}$$

for any Y tangent to M.

Let $\mathcal{N}$ be the open subset of M such that

$$\mathcal{N} = \{p \in M : \beta \neq 0, \text{ in a neighborhood of } p\}.$$

The inner product of relation (34) for $Y = \xi$ implies, taking into account relation (13), $\beta = 0$, which is impossible. Thus, $\mathcal{N}$ is empty and the following proposition has been proved

Proposition 4. *Every real hypersurface in $M_2(c)$ whose tensor field P satisfies relation (8) is a Hopf hypersurface.*

Since M is a Hopf hypersurface, Theorems 6 and 3 hold. Relation (34) for $Y = W$ implies, due to $AW = \lambda W$ and $A\phi W = \nu\phi W$,

$$(\lambda - \nu)(\nu + \lambda - 2k) = 0.$$

We have two cases:

Case I: Supposing that $\lambda \neq \nu$, then the above relation implies $\nu + \lambda = 2k$. Relation (19) implies, due to the last one, that λ, ν are constant. Thus, M is locally congruent to a real hypersurface with three distinct principal curvatures. Therefore, it is locally congruent to a real hypersurface of type (B).

Thus, in the case of $\mathbb{C}P^2$, substitution of the eigenvalues of real hypersurface of type (B) in $\nu + \lambda = 2k$ implies $\alpha = -2k$. In the case of $\mathbb{C}H^2$, substitution of the eigenvalues of real hypersurface of type (B) in $\nu + \lambda = 2k$ yields $\alpha = \frac{4}{k}$.

Case II: Supposing that $\lambda = \nu$, then the structure tensor ϕ commutes with the shape operator A, i.e., $A\phi = \phi A$ and, because of Theorem 3, M is locally congruent to a real hypersurface of type (A) and this completes the proof of Theorem 5.

As a consequence of Theorems 4 and 5, the following Corollary is obtained:

Corollary 2. *A real hypersurface M in $M_2(c)$ whose tensor field P satisfies relation (7) is locally congruent to a real hypersurface of type (A).*

5. Conclusions

In this paper, we answer the question if there are three-dimensional real hypersurfaces in non-flat complex space forms whose differential operator $\mathcal{L}^{(k)}$ of a tensor field of type (1, 1) coincides with the Lie derivative of it. First, we study the case of the tensor field being the shape operator A of the real hypersurface. The obtained results complete the work that has been done in the case of real hypersurfaces of dimensions greater than three in complex projective space (see [11]). In Table 3 all the existing results and also provides open problems are summarized.

Table 3. Results on condition $\hat{\mathcal{L}}_X^{(k)} A = \mathcal{L}_X A$.

Condition	$M_2(c)$	$\mathbb{C}P^n, n \geq 3$	$\mathbb{C}H^n, n \geq 3$
$\hat{\mathcal{L}}_X^{(k)} A = \mathcal{L}_X A, X \in \mathbb{D}$	does not exist	does not exist	open
$\hat{\mathcal{L}}_\xi^{(k)} A = \mathcal{L}_\xi A$	type (A)	type (A)	open
$\hat{\mathcal{L}}_X^{(k)} A = \mathcal{L}_X A, X \in TM$	does not exist	does not exist	open

Next, we study the above geometric condition in the case of the tensor field being $P = A\phi - \phi A$, which is introduced here. In Table 4, we summarize the obtained results.

Table 4. Results on condition $\hat{\mathcal{L}}_X^{(k)} P = \mathcal{L}_X P$.

Condition	$\mathbb{C}P^2$	$\mathbb{C}H^2$
$\hat{\mathcal{L}}_X^{(k)} P = \mathcal{L}_X P, X \in \mathbb{D}$	type (A)	type (A)
$\hat{\mathcal{L}}_\xi^{(k)} P = \mathcal{L}_\xi P$	type (A) and type (B) with $\alpha = -2k$	type (A) and type (B) with $\alpha = \frac{4}{k}$
$\hat{\mathcal{L}}_X^{(k)} P = \mathcal{L}_X P, X \in TM$	type (A)	type (A)

Author Contributions: All authors contributed equally to this research.

Acknowledgments: The authors would like to express their gratitude to the referees for valuable comments on improving the paper.

Conflicts of Interest: The authors declare no conflict of interest.

References

1. Takagi, R. On homogeneous real hypersurfaces in a complex projective space. *Osaka J. Math.* **1973**, *10*, 495–506.
2. Takagi, R. Real hypersurfaces in complex projective space with constant principal curvatures. *J. Math. Soc. Jpn.* **1975**, *27*, 43–53.
3. Takagi, R. Real hypersurfaces in complex projective space with constant principal curvatures II. *J. Math. Soc. Jpn.* **1975**, *27*, 507–516.
4. Kimura, M. Real hypersurfaces and complex submanifolds in complex projective space. *Trans. Am. Math. Soc.* **1986**, *296*, 137–149.
5. Montiel, S. Real hypersurfaces of a complex hyperbolic space. *J. Math. Soc. Jpn.* **1985**, *35*, 515–535.
6. Berndt, J. Real hypersurfaces with constant principal curvatures in complex hyperbolic space. *J. Reine Angew. Math.* **1989**, *395*, 132–141.
7. Cho, J.T. CR-structures on real hypersurfaces of a complex space form. *Publ. Math. Debr.* **1999**, *54*, 473–487.
8. Cho, J.T. Pseudo-Einstein CR-structures on real hypersurfaces in a complex space form. *Hokkaido Math. J.* **2008**, *37*, 1–17.
9. Tanno, S. Variational problems on contact Riemennian manifolds. *Trans. Am. Math. Soc.* **1989**, *314*, 349–379.
10. Pérez, J.D.; Suh, Y.J. Generalized Tanaka–Webster and covariant derivatives on a real hypersurface in a complex projective space. *Monatsh. Math.* **2015**, *177*, 637–647.
11. Pérez, J.D. Comparing Lie derivatives on real hypersurfaces in complex projective space. *Mediterr. J. Math.* **2016**, *13*, 2161–2169.
12. Panagiotidou, K.; Pérez, J.D. On the Lie derivative of real hypersurfaces in $\mathbb{C}P^2$ and $\mathbb{C}H^2$ with respect to the generalized Tanaka–Webster connection. *Bull. Korean Math. Soc.* **2015**, *52*, 1621–1630.
13. Okumura, M. On some real hypersurfaces of a complex projective space. *Trans. Am. Math. Soc.* **1975**, *212*, 355–364.
14. Montiel, S.; Romero, A. On some real hypersurfaces of a complex hyperbolic space. *Geom. Dedic.* **1986**, *20*, 245–261.
15. Panagiotidou, K.; Xenos, P.J. Real hypersurfaces in $\mathbb{C}P^2$ and $\mathbb{C}H^2$ whose structure Jacobi operator is Lie $\mathbb{D}$-parallel. *Note Mat.* **2012**, *32*, 89–99.
16. Maeda, Y. On real hypersurfaces of a complex projective space. *J. Math. Soc. Jpn.* **1976**, *28*, 529–540.
17. Ki, U.- H.; Suh, Y.J. On real hypersurfaces of a complex space form. *Math. J. Okayama Univ.* **1990**, *32*, 207–221.
18. Niebergall, R.; Ryan, P.J. Real hypersurfaces in complex space forms. In *Tight and Taut Submanifolds;* MSRI Publications: Cambridge, UK, 1997; Volume 32, pp. 233–305.

mathematics

Article

Inequalities on Sasakian Statistical Manifolds in Terms of Casorati Curvatures

Chul Woo Lee [1] and Jae Won Lee [2,*]

1 Department of Mathematics, Kyungpook National University, Daegu 41566, Korea; mathisu@knu.ac.kr
2 Department of Mathematics Education and RINS, Gyeongsang National University, Jinju 52828, Korea
* Correspondence: leejaew@gnu.ac.kr; Tel.: +82-55-772-2251

Received: 13 September 2018; Accepted: 15 November 2018; Published: 17 November 2018

Abstract: A statistical structure is considered as a generalization of a pair of a Riemannian metric and its Levi-Civita connection. With a pair of conjugate connections ∇ and ∇^* in the Sasakian statistical structure, we provide the normalized scalar curvature which is bounded above from Casorati curvatures on C-totally real (Legendrian and slant) submanifolds of a Sasakian statistical manifold of constant φ-sectional curvature. In addition, we give examples to show that the total space is a sphere.

Keywords: Sasakian statistical manifold; conjugate connection; Casorati curvature

1. Introduction

A statistical model in information geometry has a Fisher metric as a Riemannian metric with an affine connection, whose connection is constructed from the average of the probability distribution. In the statistical models, a pair of a Fisher information metric and an affine connection gives the geometric structure, called the Chentsov-Amari connection [1], whose geometric structure is a generalization of a pair of a Riemannian metric and a Levi-Civita connection. By generalizing the geometric structure, a statistical structure has been studied in information geometry. Applying this idea to Sasakian manifolds, one arrived at the definition of a Sasakian statistical structure as a generalization of a Sasakian structure. In other words, it is a triple of an affine connection, a Riemannian metric, and a Sasakian structure on an odd dimensional manifold [2]. The geometry of such a manifold is closely related to affine geometry and Hessian geometry. In such manifolds, there are the fundamental equations such as Gauss formula, Weingarten formula and the equations of Gauss, Codazzi and Ricci in submanifolds of a statistical manifold [3].

On the other hand, it is well-known that the Casorati curvature as a new extrinsic invariant is defined as the normalized square of the length of the second fundamental form, introduced by Casorati ([4,5]). Geometric meanings of Casorati curavature were found in visual perception of shape and appearance ([6–8]). Some optimal inequalities involving Casorati curvatures were proved in [9–15] for several submanifolds in real, complex and quaternionic space forms with various connections. Moreover, Lee et al. established that the normalized scalar curvature is bounded by Casorati curvatures of submanifolds in a statistical manifold of constant curvature [16]. In Kenmotsu statistical manifolds, Decu et al. investigate curvature properties and establish optimizations in terms of a new extrinsic invariant (the normalized δ-Casorati curvature) and an intrinsic invariant (the scalar curvature) [17].

In our paper, we establish optimizations of the normalized scalar curvature (the intrinsic invariant) for a new extrinsic invariant (generalized normalized Casorati curvatures) on Legendrian and slant submanifolds in a Sasakian statistical space form. Moreover, we provide some examples for special Sasakian statistical sphere S^{2m+1} of statistical sectional curvature 1.

2. Preliminaries

Let $(\overline{M}^m, \overline{g})$ be a m-dimensional Riemannian manifold with an affine connection $\overline{\nabla}$. We denote by $\Gamma(T\overline{M})$ the collection of all vector fields on $\overline{M}$.

Definition 1 ([18]). *A pair $(\overline{\nabla}, \overline{g})$ is called a statistical structure on M if $\overline{\nabla}$ is a torsion free connection on M and the covariant derivative $\overline{\nabla} g$ is symmetric.*

Definition 2. *A statistical manifold $(\overline{M}, \overline{g}, \overline{\nabla})$ is a Riemannian manifold, endowed with a pair of torsion-free affine connections $\overline{\nabla}$ and $\overline{\nabla}^*$ satisfying*

$$Z\overline{g}(X,Y) = \overline{g}(\overline{\nabla}_Z X, Y) + \overline{g}\left(X, \overline{\nabla}^*_Z Y\right) \tag{1}$$

for any vector fields X, Y and Z. The connections $\overline{\nabla}$ and $\overline{\nabla}^*$ are called dual connections.

Remark 1.

(a) $\left(\overline{\nabla}^*\right)^* = \overline{\nabla}$.

(b) *If $(\overline{\nabla}, \overline{g})$ is a statistical structure, then so is $\left(\overline{\nabla}^*, g\right)$.*

(c) *Any torsion-free affine connection $\overline{\nabla}$ always has a dual connection satisfying*

$$\overline{\nabla} + \overline{\nabla}^* = 2\overline{\nabla}^0, \tag{2}$$

where $\overline{\nabla}^0$ is the Levi-Civita connection for $\overline{M}$.

Let $\overline{R}$ and $\overline{R}^*$ be the curvature tensor fields of $\overline{\nabla}$ and $\overline{\nabla}^*$, respectively.

Definition 3 ([18,19]). *Let $(\overline{\nabla}, \overline{g})$ be a statistical structure on $\overline{M}$. We define*

$$S(X,Y)Z = \frac{1}{2}\{\overline{R}(X,Y)Z + \overline{R}^*(X,Y)Z\}$$

for $X, Y, Z \in \Gamma(T\overline{M})$, called the statistical curvature tensor of $(\overline{\nabla}, \overline{g})$. In particular, a statistical manifold $(\overline{M}, \overline{\nabla}, \overline{g})$ is to be of constant statistical curvature $c \in \mathbb{R}$ if $S(X,Y)Z = c\{\overline{g}(Y,Z)X - \overline{g}(X,Z)Y\}$ for $X, Y, Z \in \Gamma(T\overline{M})$.

By the direct calculation, the curvature tensor fields $\overline{R}$ and $\overline{R}^*$ satisfy

$$\overline{g}\left(\overline{R}^*(X,Y)Z, W\right) = -\overline{g}(Z, \overline{R}(X,Y)W), \quad X, Y, Z, W \in \Gamma(T\overline{M}).$$

Therefore, if $(\overline{\nabla}, \overline{g})$ is a statistical structure of constant curvature c, so is $\left(\overline{\nabla}^*, \overline{g}\right)$.

For submanifolds in statistical manifolds, we have pairs of induced connections ∇, ∇^*, second fundamental forms h, h^*, shape operators A, A^*, and normal connections D, D^* satisfying equations analogous to the Gauss and the Weingarten ones for $\overline{\nabla}$ and $\overline{\nabla}^*$, respectively. Moreover, the induced metric g is unique, and (∇, g) and (∇^*, g) are induced dual statistical structures on the submanifold. The fundamental equations for statistical submanifolds are given by Vos ([3]).

Let (M, g) be an n-dimensional submanifold of a statistical manifold $(\overline{M}, \overline{g})$ and g the induced metric on M. Then for any vector fields X, Y, the Gauss formulas are given respectively by

$$\begin{aligned}\overline{\nabla}_X Y &= \nabla_X Y + h(X,Y)\\ \overline{\nabla}^*_X Y &= \nabla^*_X Y + h^*(X,Y).\end{aligned}$$

The corresponding Gauss equations with respect to $\overline{\nabla}$ and $\overline{\nabla}^*$ are given by the following result.

Theorem 1 ([3]). *Let $\overline{\nabla}$ and $\overline{\nabla}^*$ be dual connections on $(\overline{M},\overline{g})$ and ∇ and ∇^* the induced dual connections by $\overline{\nabla}$ and $\overline{\nabla}^*$ by a submanifold M of $(\overline{M},\overline{g})$, respectively. Let $\overline{R}$, R, $\overline{R^*}$ and R^* be the Riemannian curvature tensors of $\overline{\nabla}$, ∇, $\overline{\nabla^*}$ and ∇^*, respectively. Then*

$$\begin{aligned}\overline{g}\left(\overline{R}\left(X,Y\right)Z,W\right) &= \overline{g}\left(R\left(X,Y\right)Z,W\right) \\ &\quad + \overline{g}\left(h\left(X,Z\right),h^*\left(Y,W\right)\right) - \overline{g}\left(h^*\left(X,W\right),h\left(Y,Z\right)\right)\end{aligned} \tag{3}$$

$$\begin{aligned}\overline{g}\left(\overline{R}^*\left(X,Y\right)Z,W\right) &= \overline{g}\left(R^*\left(X,Y\right)Z,W\right) \\ &\quad + \overline{g}\left(h^*\left(X,Z\right),h\left(Y,W\right)\right) - \overline{g}\left(h\left(X,W\right),h^*\left(Y,Z\right)\right)\end{aligned} \tag{4}$$

If $\{e_1,...,e_n\}$ is an orthonormal basis of the tangent space T_pM and $\{e_{n+1},...,e_m\}$ is an orthonormal basis of the normal space $T_p^{\perp}M$, then the scalar curvature τ at p is defined as

$$\tau(p) = \sum_{1\le i<j\le n} g\left(S\left(e_i,e_j\right)e_j,e_i\right)$$

and the normalized scalar curvature ρ of M is defined as

$$\rho = \frac{2\tau}{n(n-1)}.$$

We denote by H, H^* the mean curvature vectors, that is,

$$H(p) = \frac{1}{n}\sum_{i=1}^{n} h(e_i,e_i), \qquad H^*(p) = \frac{1}{n}\sum_{i=1}^{n} h^*(e_i,e_i) \tag{5}$$

and we also set

$$h_{ij}^{\alpha} = g(h(e_i,e_j),e_\alpha), \quad h_{ij}^{*\alpha} = g(h^*(e_i,e_j),e_\alpha),$$

$i,j \in \{1,...,n\}$, $\alpha \in \{n+1,...,m\}$.

Then it is well-known that the squared mean curvatures of the submanifold M in $\overline{M}$ are defined by

$$\|H\|^2 = \frac{1}{n^2}\sum_{\alpha=n+1}^{m}\left(\sum_{i=1}^{n} h_{ii}^{\alpha}\right)^2, \quad \|H^*\|^2 = \frac{1}{n^2}\sum_{\alpha=n+1}^{m}\left(\sum_{i=1}^{n} h_{ii}^{*\alpha}\right)^2$$

and the squared norms of h and h^* over dimension n is denoted by $\mathcal{C}$ and $\mathcal{C}^*$ are called the Casorati curvatures of the submanifold M, respectively. Therefore, we have

$$\mathcal{C} = \frac{1}{n}\sum_{\alpha=n+1}^{m}\sum_{i,j=1}^{n}\left(h_{ij}^{\alpha}\right)^2 \text{ and } \mathcal{C}^* = \frac{1}{n}\sum_{\alpha=n+1}^{m}\sum_{i,j=1}^{n}\left(h_{ij}^{*\alpha}\right)^2.$$

The normalized δ-Casorati curvatures $\delta_C(n-1)$ and $\widehat{\delta}_C(n-1)$ of the submanifold M are defined as

$$[\delta_C(n-1)]_p = \frac{1}{2}\mathcal{C}_p + \frac{(n+1)}{2n}\inf\{\mathcal{C}(L)|L \text{ a hyperplane of } T_pM\},$$

and

$$\left[\widehat{\delta}_C(n-1)\right]_p = 2\mathcal{C}_p - \frac{(2n-1)}{2n}\sup\{\mathcal{C}(L)|L \text{ a hyperplane of } T_pM\}.$$

Similarly, the dual normalized δ^*-Casorati curvatures $\delta_C^*(n-1)$ and $\widehat{\delta}_C^*(n-1)$ of the submanifold M are defined as

$$[\delta_C^*(n-1)]_p = \frac{1}{2}C_p^* + \frac{(n+1)}{2n}\inf\{C^*(L)|L \text{ a hyperplane of } T_pM\},$$

and

$$\left[\widehat{\delta}_C^*(n-1)\right]_p = 2C_p^* - \frac{(2n-1)}{2n}\sup\{C^*(L)|L \text{ a hyperplane of } T_pM\}.$$

The generalized normalized δ-Casorati curvatures $\delta_C(t;n-1)$ and $\widehat{\delta}_C(t;n-1)$ of the submanifold M are defined for any positive real number $t \neq n(n-1)$ as

$$[\delta_C(t;n-1)]_p = tC_p + \frac{(n-1)(n+t)(n^2-n-t)}{nt}\inf\{C(L)|L \text{ a hyperplane of } T_pM\},$$

if $0 < t < n^2 - n$, and

$$\left[\widehat{\delta}_C(t;n-1)\right]_p = tC_p - \frac{(n-1)(n+t)(t-n^2+n)}{nt}\sup\{C(L)|L \text{ a hyperplane of } T_pM\},$$

if $t > n^2 - n$.

Moreover, the dual generalized normalized δ-Casorati curvatures $\delta_C^*(t;n-1)$ and $\widehat{\delta}^*{}_C(t;n-1)$ of the submanifold M are defined for any positive real number $t \neq n(n-1)$ as

$$[\delta_C^*(t;n-1)]_p = tC_p^* + \frac{(n-1)(n+t)(n^2-n-t)}{nt}\inf\{C^*(L)|L \text{ a hyperplane of } T_pM\},$$

if $0 < t < n^2 - n$, and

$$\left[\widehat{\delta}^*{}_C(t;n-1)\right]_p = tC_p^* - \frac{(n-1)(n+t)(t-n^2+n)}{nt}\sup\{C^*(L)|L \text{ a hyperplane of } T_pM\},$$

if $t > n^2 - n$.

The following lemma plays a key role in the proof of our main theorem.

Lemma 1 ([20]). *Let*

$$\Gamma = \{(x_1, x_2, \cdots, x_n) \in \mathbb{R}^n : x_1 + x_2 + \cdots + x_n = k\}$$

be a hyperplane of $\mathbb{R}^n$*, and* $f : \mathbb{R}^n \longrightarrow \mathbb{R}$ *a quadratic form, given by*

$$f(x_1, x_2, \cdots, x_n) = a\sum_{i=1}^{n-1}(x_i)^2 + b\,(x_n)^2 - 2\sum_{1\le i<j\le n} x_i x_j, \qquad a > 0,\ b > 0.$$

Then, the constrained extremum problem $\min_{x\in\Gamma} f(x)$ *has a global solution as follows:*

$$x_1 = x_2 = \cdots = x_{n-1} = \frac{k}{a+1}, \quad x_n = \frac{k}{b+1} = \frac{k(n-1)}{(a+1)b} = (a-n+2)\frac{k}{a+1},$$

provided that

$$b = \frac{n-1}{a-n+2}.$$

Definition 4. *A triple $(\overline{g}, \varphi, \xi)$ is called an almost contact metric structure on $\overline{M}$ if the following equations hold*

$$\varphi\xi = 0, \quad \overline{g}(\xi, \xi) = 1, \quad \varphi^2 X = -X + \overline{g}(X, \xi)\xi, \quad \overline{g}(\varphi X, Y) + \overline{g}(X, \varphi Y) = 0, \quad X, Y \in \Gamma(T\overline{M})$$

where φ is a section of $T\overline{M} \otimes T\overline{M}^*$ and ξ is the structure vector field on $\overline{M}$.

Definition 5. *A quadraple $(\overline{\nabla}, \overline{g}, \varphi, \xi)$ is called a Sasakian statistical structure on $\overline{M}$ if $(\overline{\nabla}, \overline{g})$ is a statistical structure.*

Theorem 2 ([2]). *Let $(\overline{\nabla}, \overline{g}, \varphi, \xi)$ be a Sasakian statistical structure on $\overline{M}$. Then, so is $\left(\overline{\nabla}^*, \overline{g}, \varphi, \xi\right)$.*

Definition 6. *Let $(\overline{\nabla}, \overline{g}, \varphi, \xi)$ be a Sasakian statistical structure on $\overline{M}$, and $c \in \mathbb{R}$. The Sasakian statistical structure is said to be of constant φ-sectional curvature if*

$$\begin{aligned} S(X,Y)Z = {} & \frac{c+3}{4}\{\overline{g}(Y,Z)X - \overline{g}(X,Z)Y\} + \frac{c-1}{4}\{\overline{g}(\varphi Y, Z)\varphi X \\ & - \overline{g}(\varphi X, Z)\varphi Y - 2\overline{g}(\varphi X, Y)\varphi Z - \overline{g}(Y,\xi)\overline{g}(Z,\xi)X \\ & + \overline{g}(X,\xi)\overline{g}(Z,\xi)Y + \overline{g}(Y,\xi)\overline{g}(Z,X)\xi - \overline{g}(X,\xi)\overline{g}(Z,Y)\xi\}, \end{aligned} \tag{6}$$

$X, Y, Z \in \Gamma(T\overline{M})$.

A submanifold M^n normal to ξ in a Sasakian statistical manifold $\overline{M}^{2m+1}$ is said to be a *C-totally real* submanifold. In this case, $\varphi\left(T_pM\right) \subset T_p^{\perp}M$, $p \in M$. In particular, if $n = m$, then M is called a *Legendrian* submanifold.

For submanifolds tangent to ξ, there is a θ-slant submanifold of a Sasakian statistical manifold as follows [21]:

A submanifold M^n tangent to ξ in a Sasakian statistical manifold is called a *θ-slant submanifold* if for any vector $X \in T_pM$, linearly independent on ξ_p, the angle between φX and T_pM is a constant $\theta \in [0, \frac{\pi}{2}]$, called the slant angle of M in $\overline{M}$. In particular, if $\theta = 0$ and $\theta = \frac{\pi}{2}$, M is invariant and anti-invariant, respectively.

3. Inequalities with Casorati Curvatures

Let M be an n-dimensional C-totally real submanifold of a $(2m+1)$-dimensional Sasakian statistical manifold $(\overline{M}, \overline{\nabla}, \overline{g}, \varphi, \xi)$.

Let $p \in M$ and the set $\{e_1, e_2, \cdots, e_n\}$ and $\{e_{n+1}, e_{n+2}, \cdots, e_{2m}, e_{2m+1} = \xi\}$ be orthonormal bases of T_pM and $T_p^{\perp}M$, respectively. Then, we have the scalar curvature as follows:

$$\begin{aligned} 2\tau(p) &= 2\sum_{1\le i<j\le n} g\left(S\left(e_i, e_j\right)e_j, e_i\right) \\ &= \sum_{1\le i<j\le n} \{\overline{g}\left(R(e_i,e_j)e_j, e_i\right) + \overline{g}\left(R^*(e_i,e_j)e_j, e_i\right)\} \\ &= \sum_{1\le i<j\le n} \{\frac{c+3}{2} + \overline{g}\left(h(e_i,e_i), h^*(e_j,e_j)\right) + \overline{g}\left(h^*(e_i,e_i), h(e_j,e_j)\right) \\ &\quad - 2\overline{g}\left(h^*(e_i,e_j), h(e_i,e_j)\right)\} \\ &= \frac{n(n-1)(c+3)}{4} + n^2\overline{g}(H, H^*) - \sum_{i,j=1}^{n} \overline{g}\left(h^*(e_i,e_j), h(e_i,e_j)\right) \end{aligned} \tag{7}$$

Since $2H^0 = H + H^*$ and the definition of Casorati curvature, $4\|H^0\|^2 = \|H\|^2 + \|H^*\|^2 + 2g(H, H^*)$, we obtain that

$$2\tau(p) = \frac{n(n-1)(c+3)}{4} + 2n^2\|H^0\|^2 - \frac{n^2}{2}\left(\|H\|^2 + \|H^*\|^2\right) - 2n\mathcal{C}^0 + \frac{n}{2}(\mathcal{C} + \mathcal{C}^*), \tag{8}$$

where $\mathcal{C}^0 = \frac{1}{2}(\mathcal{C} + \mathcal{C}^*)$.

Define a quadratic polynomial in the components of the second fundamental form h^0 by

$$\mathcal{P} = t\mathcal{C}^0 + \frac{(n-1)(n+t)(n^2-n-t)}{nt}\mathcal{C}^0(L) + \frac{1}{2}n(\mathcal{C} + \mathcal{C}^*) - \frac{n^2}{2}\left(\|H\|^2 + \|H^*\|^2\right) - 2\tau(p) + \frac{n(n-1)(c+3)}{4},$$

where L is a hyperplane of T_pM. Without loss of generality, we can assume that L is spanned by $e_1, \cdots, e_{n-1}$. Then we derive

$$\begin{aligned}\frac{1}{2}\mathcal{P} =& \sum_{\alpha=n+1}^{m}\sum_{i=1}^{n-1}\left[\frac{n^2+n(t-1)-2t}{r}\left(h_{ii}^{0\alpha}\right)^2 + \frac{2(n+t)}{n}\left(h_{in}^{0\alpha}\right)^2\right] \\ &+ \sum_{\alpha=n+1}^{m}\left[\frac{2(n+t)(n-1)}{t}\sum_{1=i<j}^{n-1}\left(h_{ij}^{0\alpha}\right)^2 - 2\sum_{1=i<j}^{n} h_{ii}^{0\alpha}h_{jj}^{0\alpha} + \frac{t}{n}\left(h_{nn}^{0\alpha}\right)^2\right] \\ \geq& \sum_{\alpha=n+1}^{m}\left[\sum_{i=1}^{n-1}\frac{n^2+n(t-1)-2t}{t}\left(h_{ii}^{0\alpha}\right)^2 - 2\sum_{1=i<j}^{n} h_{ii}^{0\alpha}h_{jj}^{0\alpha} + \frac{t}{n}\left(h_{nn}^{0\alpha}\right)^2\right].\end{aligned} \tag{9}$$

For $\alpha = n+1, \cdots, m$, let us consider the quadratic form $f_\alpha : \mathbb{R}^n \longrightarrow \mathbb{R}$ defined by

$$f_\alpha\left(h_{11}^{0\alpha}, \cdots, h_{nn}^{0\alpha}\right) = \frac{n^2+n(t-1)-2t}{t}\sum_{i=1}^{n-1}\left(h_{ii}^{0\alpha}\right)^2 - 2\sum_{1=i<j}^{n} h_{ii}^{0\alpha}h_{jj}^{0\alpha} + \frac{t}{n}\left(h_{nn}^{0\alpha}\right)^2, \tag{10}$$

and the constrained extremum problem

$$\min f_\alpha$$

$$\text{subject to } F^\alpha : h_{11}^{0\alpha} + \cdots + h_{nn}^{0\alpha} = c^\alpha,$$

where c^α is a real constant. Comparing (10) with the quadratic function in Lemma 1, we see that

$$a = \frac{n^2+n(t-1)-2t}{t}, \qquad b = \frac{t}{n}.$$

Therefore, we have the critical point $(h_{11}^{0\alpha}, \cdots, h_{nn}^{0\alpha})$, given by

$$h_{11}^{0\alpha} = h_{22}^{0\alpha} = \cdots = h_{n-1\,n-1}^{0\alpha} = \frac{tc^\alpha}{(n+t)(n-1)}, \qquad h_{nn}^{0\alpha} = \frac{nc^\alpha}{n+t},$$

is a global minimum point by Lemma 1. Moreover, $f_\alpha\left(h_{11}^{0\alpha}, \cdots, h_{nn}^{0\alpha}\right) = 0$. Therefore, we have

$$\mathcal{P} \geq 0, \tag{11}$$

which implies

$$2\tau(p) \leq t\mathcal{C}^0 + \frac{(n-1)(n+t)(n^2-n-t)}{nt}\mathcal{C}^0(L) + \frac{1}{2}n(\mathcal{C} + \mathcal{C}^*) - \frac{n^2}{2}\left(\|H\|^2 + \|H^*\|^2\right) + \frac{n(n-1)(c+3)}{4}.$$

Therefore, we derive

$$\rho \leq \frac{1}{n(n-1)}\{tC^0 + \frac{(n-1)(n+t)(n^2-n-t)}{nt}C^0(L)\}$$
$$+\frac{1}{2(n-1)}(C+C^*) - \frac{n}{2(n-1)}\left(\|H\|^2 + \|H^*\|^2\right) + \frac{c+3}{4}.$$

Therefore, we have the following theorem:

Theorem 3. *Let M be an n-dimensional C-totally real submanifold of a $(2m+1)$-dimensional Sasakian statistical manifold $(\overline{M}, \overline{\nabla}, \overline{g}, \varphi, \xi)$. When $0 < t < n^2 - n$, the generalized normalized δ-Casorati curvature $\delta_C^0(t, n-1)$ on M satisfies*

$$\rho \leq \frac{1}{n(n-1)}\delta_C^0(t, n-1) + \frac{1}{2(n-1)}(C+C^*)$$
$$-\frac{n}{2(n-1)}\left(\|H\|^2 + \|H^*\|^2\right) + \frac{c+3}{4},$$

where $2\delta_C^0(t, n-1) = \delta_C(t, n-1) + \delta_C^*(t, n-1)$. The equality case holds identically at any point $p \in M$ if and only if $h = -h^*$.

For a unit hypersphere S^{2n+1} in $\mathbb{R}^{2n+2}$, the unit normal vector field N of S^{2n+1} provides the structure vector field $\xi = -JN$ with the standard almost complex structure J on $\mathbb{R}^{2n+2} = \mathbb{C}^{n+1}$. In addition, $\varphi = \pi \circ J$ is the natural projection of the tangent space of $\mathbb{R}^{2n+2}$ onto the tangent space of S^{2n+1}. Then we obtain the standard Sasakian structure (g, φ, ξ) on S^{2n+1}. From [2], we can construct a Sasakian statistical structures on S^{2n+1} of constant statistical sectional curvature 1. Therefore, we have the following optimal inequality:

Example 1. *Let M be an n-dimensional C-totally real submanifold of S^{2m+1}. Then, the generalized normalized δ-Casorati curvature $\delta_C^0(t, n-1)$ on M^n satisfies*

$$\rho \leq \frac{1}{n(n-1)}\delta_C^0(t, n-1) + \frac{1}{2(n-1)}(C+C^*)$$
$$-\frac{n}{2(n-1)}\left(\|H\|^2 + \|H^*\|^2\right) + 1.$$

When $t = \frac{n(n-1)}{2}$ in Theorem 3, we have an optimization for a normalized δ-Casoratic curvature as follows:

Corollary 1. *Let M be an n-dimensional C-totally real submanifold of a $(2m+1)$-dimensional Sasakian statistical manifold $(\overline{M}, \overline{\nabla}, \overline{g}, \varphi, \xi)$. Then, the normalized δ-Casorati curvature $\delta_C^0(n-1)$ on M satisfies*

$$\rho \leq \delta_C^0(n-1) + \frac{1}{2(n-1)}(C+C^*) - \frac{n}{2(n-1)}\left(\|H\|^2 + \|H^*\|^2\right) + \frac{c+3}{4}.$$

Proof. Taking $t = \frac{n(n-1)}{2}$ in $\delta_C^0(t, n-1)$, we have the following relation:

$$\left[\delta_C^0\left(\frac{n(n-1)}{2}; n-1\right)\right]_p = n(n-1)\left[\delta_C^0(n-1)\right]_p$$

in any point $p \in M$. Therefore, we have an optimal inequality for the normalized δ-Casorati curvature $\delta_C^0(n-1)$. □

Theorem 4. *Let M be an n-dimensional θ-slant submanifold of a $(2m+1)$-dimensional Sasakian statistical manifold $(\overline{M}, \overline{\nabla}, \overline{g}, \varphi, \xi)$. When $0 < t < n^2 - n$, the generalized normalized δ-Casorati curvature $\delta_C^0(t, n-1)$ on M satisfies*

$$\rho \leq \frac{1}{n(n-1)}\delta_C^0(t,n-1) + \frac{1}{2(n-1)}\left(\mathcal{C}+\mathcal{C}^*\right) - \frac{n}{2(n-1)}\left(\|H\|^2 + \|H^*\|^2\right) + \frac{n(n-1)(c+3)}{4} + \frac{3(n-1)(c-1)\cos^2\theta}{4} - \frac{(n-1)(c-1)}{2}.$$

Proof. Let $p \in M$ and the set $\{e_1, e_2, \cdots, e_{n-1}, e_n = \xi\}$ and $\{e_{n+1}, e_{n+2}, \cdots, e_{2m}, e_{2m+1}\}$ be orthonormal bases of T_pM and $T_p^{\perp}M$, respectively. Then, we have the scalar curvature as follows:

$$\begin{aligned} 2\tau(p) &= 2\sum_{1\leq i<j\leq n} g\left(S\left(e_i, e_j\right)e_j, e_i\right) \\ &= \sum_{1\leq i<j\leq n}\left\{\overline{g}\left(R(e_i,e_j)e_j,e_i\right) + \overline{g}\left(R^*(e_i,e_j)e_j,e_i\right)\right\} \\ &= \frac{n(n-1)(c+3)}{4} + \frac{3(n-1)(c-1)\cos^2\theta}{4} - \frac{(n-1)(c-1)}{2} \\ &+ n^2\overline{g}(H,H^*) - \sum_{i,j=1}^{n}\overline{g}\left(h^*(e_i,e_j), h(e_i,e_j)\right) \end{aligned} \tag{12}$$

By using a similar argument as in the proof of Theorem 3, we get

$$\begin{aligned} 2\tau(p) \leq & t\mathcal{C}^0 + \frac{(n-1)(n+t)(n^2-n-t)}{nt}\mathcal{C}^0(L) \\ &+ \frac{1}{2}n\left(\mathcal{C}+\mathcal{C}^*\right) - \frac{n^2}{2}\left(\|H\|^2 + \|H^*\|^2\right) \\ &+ \frac{n(n-1)(c+3)}{4} + \frac{3(n-1)(c-1)\cos^2\theta}{4} - \frac{(n-1)(c-1)}{2}. \end{aligned}$$

Therefore, we have an ineqaulity as follows:

$$\rho \leq \frac{1}{n(n-1)}\delta_C^0(t,n-1) + \frac{1}{2(n-1)}\left(\mathcal{C}+\mathcal{C}^*\right) - \frac{n}{2(n-1)}\left(\|H\|^2 + \|H^*\|^2\right) + \frac{n(n-1)(c+3)}{4} + \frac{3(n-1)(c-1)\cos^2\theta}{4} - \frac{(n-1)(c-1)}{2}.$$

□

If M is an invariant submanifold, then $\theta = 0$. Then we obtain

Corollary 2. *Let M^n be an n-dimensional invariant submanifold of a $(2m+1)$-dimensional Sasakian statistical manifold $(\overline{M}, \overline{\nabla}, \overline{g}, \varphi, \xi)$. When $0 < t < n^2 - n$, we derive*

$$\rho \leq \frac{1}{n(n-1)}\delta_C^0(t,n-1) + \frac{1}{2(n-1)}\left(\mathcal{C}+\mathcal{C}^*\right) - \frac{n}{2(n-1)}\left(\|H\|^2 + \|H^*\|^2\right) + \frac{n(n-1)(c+3)}{4} + \frac{(n-1)(c-1)}{4}.$$

If M is an anti-invariant submanifold, then $\theta = \frac{\pi}{2}$. Then we obtain

Corollary 3. *Let M^n be an n-dimensional anti-invariant submanifold of a $(2m+1)$-dimensional Sasakian statistical manifold $(\overline{M}, \overline{\nabla}, \overline{g}, \varphi, \xi)$. When $0 < t < n^2 - n$, we derive*

$$\rho \leq \frac{1}{n(n-1)}\delta^0_{\mathcal{C}}(t, n-1) + \frac{1}{2(n-1)}(\mathcal{C} + \mathcal{C}^*) - \frac{n}{2(n-1)}\left(\|H\|^2 + \|H^*\|^2\right) + \frac{n(n-1)(c+3)}{4} - \frac{(n-1)(c-1)}{2}.$$

Example 2. *Let M be an n-dimensional θ-slant submanifold of S^{2m+1}. Then, the generalized normalized δ-Casorati curvature $\delta^0_{\mathcal{C}}(t, n-1)$ on M^n satisfies*

$$\rho \leq \frac{1}{n(n-1)}\delta^0_{\mathcal{C}}(t, n-1) + \frac{1}{2(n-1)}(\mathcal{C} + \mathcal{C}^*) - \frac{n}{2(n-1)}\left(\|H\|^2 + \|H^*\|^2\right) + n(n-1).$$

Remark 2.

(1) *Taking $t = \frac{n(n-1)}{2}$ as Corollary 1, we have optimal inequalities for θ-slant submanifold of a Sasakian statistical manifold.*

(2) *In any optimization throughout our paper, the equality cases hold if and only if a submanifold is totally geodesic from $h = -h^*$.*

(3) *In the case for $t > n^2 - n$, the methods of finding the above inequalities are analogous.*

Author Contributions: C.W.L. presented the idea to establish optimizations on C-totally real (Legendrian, slant) submanifolds. J.L. checked and polished the draft.

Acknowledgments: Chul Woo Lee was supported by Basic Science Research Program through the National Research Foundation of Korea (NRF) funded by the Ministry of Education (2018R1D1A1B07040576) and Jae Won Lee was supported by the Basic Science Research Program through the National Research Foundation of Korea (NRF) funded by the Ministry of Education (2017R1D1A1B03033978).

Conflicts of Interest: The authors declare no conflict of interest.

References

1. Amari, S. *Differential-Geometrical Methods in Statistics, Lecture Notes in Statistics;* Springer: New York, NY, USA, 1985.
2. Furuhata, H.; Hasegawa, I.; Okuyama, Y.; Sato, K.; Shahid, M. Sasakian statistical manifolds. *J. Geom. Phys.* **2017**, *117*, 179–186. [CrossRef]
3. Vos, P.W. Fundamental equations for statistical submanifolds with applications to the Bartlett correction. *Ann. Inst. Stat. Math.* **1989**, *41*, 429–450. [CrossRef]
4. Casorati, F. Mesure de la courbure des surfaces suivant l'idée commune. Ses rapports avec les mesures de courbure gaussienne et moyenne. *Acta Math.* **1890**, *14*, 95–110. [CrossRef]
5. Kowalczyk, D. Casorati curvatures. *Bull. Transilv. Univ. Brasov Ser. III* **2008**, *1*, 209–213.
6. Koenderink, J.J. Surface shape. In *The Science and the Looks in "Handbook of Experimental Phenomenology: Visual Perception of Shape, Space and Appearance"*; Albertazzi, L., Ed.; John Wiley and Sons Ltd: Chichester, UK, 2013; pp. 165–180.
7. Ons, B.; Verstraelen, L. Some geometrical comments on vision and neurobiology: Seeing Gauss and Gabor walking by, when looking through the window of the Parma at Leuven in the company of Casorati. *Kragujevac J. Math.* **2011**, *35*, 317–325.
8. Verstraelen, L. Geometry of submanifolds I. The first Casorati curvature indicatrices. *Kragujevac J. Math.* **2013**, *37*, 5–23.
9. Decu, S.; Haesen, S.; Verstraelen, L. Optimal inequalities involving Casorati curvatures. *Bull. Transilv. Univ. Braşov Ser. B (N.S.)* **2007**, *14*, 85–93.

10. Decu, S.; Haesen, S.; Verstraelen, L. Optimal inequalities characterising quasi-umbilical submanifolds. *J. Inequal. Pure Appl. Math.* **2008**, *9*, 1–7.
11. Lee, C.W.; Yoon, D.W.; Lee, J.W. Optimal inequalities for the Casorati curvatures of submanifolds of real space forms endowed with semi-symmetric metric connections. *J. Inequal. Appl.* **2014**, *2014*. [CrossRef]
12. Lee, J.W.; Vîlcu, G.-E. Inequalities for generalized normalized δ-Casorati curvatures of slant submanifolds in quaternionic space forms. *Taiwanese J. Math.* **2015**, *19*, 691–702. [CrossRef]
13. Lee, C.W.; Lee, J.W.; Vîlcu, G.-E.; Yoon, D.W. Optimal inequalities for the Casorati curvatures of submanifolds of generalized space forms endowed with semi-symmetric metric connections. *Bull. Korean Math. Soc.* **2015**, *52*, 1631–1647. [CrossRef]
14. Lee, J.W.; Lee, C.W.; Yoon, D.W. Inequalities for generalized δ-Casorati curvatures of submanifolds in real space forms endowed with a semi-symmetric metric connection. *Rev. Union Mat. Argent.* **2016**, *57*, 53–62.
15. Slesar, V.; Şahin, B.; Vîlcu, G.-E. Inequalities for the Casorati curvatures of slant submanifolds in quaternionic space forms. *J. Inequal. Appl.* **2014**, *2014*. [CrossRef]
16. Lee, C.W.; Yoon, D.W.; Lee, J.W. A pinching theorem for statistical manifolds with Casorati curvatures. *J. Nonlinear Sci. Appl.* **2017**, *10*, 4908–4914. [CrossRef]
17. Decu, S.; Haesen, S.; Verstraelan, L.; Vilcu, G.-E. Curvature invariants of statistical submanifolds in Kenmotsu statistical manifolds of constant ϕ-sectional curvature. *Entropy* **2018**, *20*, 529. [CrossRef]
18. Opozda, B. A sectional curvature for statistical structures. *Linear Algebra Appl.* **2016**, *497*, 134–161. [CrossRef]
19. Opozda, B. Bochner's technique for statistical structures. *Ann. Glob. Anal. Geom.* **2015**, *48*, 357–395. [CrossRef]
20. Tripathi, M.M. Inequalities for algebraic Casorati curvatures and their applications. *Note Mater.* **2017**, *37*, 161–186.
21. Lotta, A. Slant submanifolds in contact geometry. *Bull. Math. Soc. Roum.* **1996**, *39*, 183–198.

Σ mathematics

MDPI

Article

Pinching Theorems for a Vanishing C-Bochner Curvature Tensor

Jae Won Lee [1] and Chul Woo Lee [2,*]

[1] Department of Mathematics Education and RINS, Gyeongsang National University, Jinju 52828, Korea; leejaew@gnu.ac.kr
[2] Department of Mathematics, Kyungpook National University, Daegu 41566, Korea
* Correspondence: mathisu@knu.ac.kr; Tel.: +82-10-3396-5292

Received: 21 September 2018; Accepted: 26 October 2018; Published: 30 October 2018

Abstract: The main purpose of this article is to construct inequalities between a main intrinsic invariant (the normalized scalar curvature) and an extrinsic invariant (the Casorati curvature) for some submanifolds in a Sasakian manifold with a zero C-Bochner tensor.

Keywords: C-Bochner tensor; generalized normalized δ-Casorati curvature; Sasakian manifold; slant; invariant; anti-invariant

1. Introduction

Bochner [1] introduced the Bochner tensor in Kähler manifolds by analogy to the Weyl conformal curvature tensor. The Bochner tensor is equal to the 4-th order Chern–Moser curvature tensor in CR-manifolds by Webster [2]. In contact manifolds, the Bochner tensor was reinterpreted by Matsumoto and Chuman [3] as a C-Bochner curvature tensor in Sasakian manifolds. They showed that a Sasakian space form is a space with a vanishing C-Bochner curvature tensor. A Sasakian manifold with a non-constant φ-sectional curvature and a vanishing C-Bochner curvature tensor was constructed by Kim [4]. Tano showed that the C-Bochner curvature tensor is invariant in terms of D-homothetic deformations [5].

On the other hand, F. Casorati introduced a new extrinsic invariant of submanifolds in a Riemannian manifold, called the Casorati curvature. This curvature is defined as the normalized square of the length of the second fundamental form ([6,7]). Moreover, there are very interesting optimizations involving Casorati curvatures, proved in [8–19] for various basic submanifolds in different spaces (real, complex, and quaternionic space forms) with several connections.

In our paper, we investigate new optimal inequalities involving Casorati curvatures for some submanifolds of a Sasakian manifold with a zero C-Bochner curvature tensor and characterize those submanifolds for which the equalities hold.

2. Preliminaries

In this section, we recall some results on almost contact manifolds and give a brief review of basic facts of C-Bochner curvature tensor.

A manifold $\overline{M} = (\overline{M}, \varphi, \xi, \eta, \overline{g})$ is called *an almost contact metric manifold* if there exist structure tensors $(\varphi, \xi, \eta, \overline{g})$, where φ is a tensor field of type $(1,1)$, ξ is a vector field, η is a 1-form, and $\overline{g}$ is the Riemannian metric on $\overline{M}$ satisfying [20]

$$\varphi\xi = 0, \quad \eta \circ \varphi = 0, \quad \eta(\xi) = 1$$

$$\varphi^2 = -I + \eta \otimes \xi, \quad \text{and} \quad \overline{g}(\varphi X, \varphi Y) = \overline{g}(X, Y) - \eta(X)\eta(Y)$$

where $I : T\overline{M} \longrightarrow T\overline{M}$ is the identity endomorphism, and X, Y are vector fields on $\overline{M}$. In particular, if $\overline{M}$ is Sasakian [21], then we have

$$(\nabla_X \varphi)Y = -\overline{g}(X,Y)\xi + \eta(Y)X \quad \text{and} \quad \overline{\nabla}_X \xi = \varphi X$$

where $\overline{\nabla}$ is the Levi–Civita connection on $\overline{M}$.

Let M^n be an n-dimensional submanifold of a Riemannian manifold $(\overline{M}, \overline{g})$. If ∇ is the induced covariant differentiation on M of the Levi–Civita connection $\overline{\nabla}$ on $\overline{M}$, then we have the Gauss and Weingarten formulas:

$$\overline{\nabla}_X Y = \nabla_X Y + h(X,Y) \forall X, Y \in \Gamma(TM)$$

and

$$\overline{\nabla}_X N = -A_N X + \nabla^{\perp}_X N, \forall X \in \Gamma(TM), \forall N \in \Gamma(T^{\perp}M)$$

where h is the second fundamental form of M, $\nabla^{\perp}$ is the connection on $T^{\perp}M$, and A_N is the shape operator of M with respect to a normal section N. If we denote by $\overline{R}$ and R the curvature tensor fields of $\overline{\nabla}$ and ∇, respectively, then we have the Gauss equation:

$$\begin{aligned} \overline{R}(X,Y,Z,W) = R(X,Y,Z,W) + \overline{g}(h(X,W), h(Y,Z)) \\ - \overline{g}(h(X,Z), h(Y,W)) \end{aligned} \tag{1}$$

for all $X, Y, Z, W \in \Gamma(TM)$.

Let M^n be an n-dimensional Riemannian submanifold of a Sasakian manifold $(\overline{M}, \overline{g}, \varphi, \xi, \eta)$. A plane section $\pi \subset T_pM$, $p \in M$ of a Sasakian manifold $\overline{M}$ is called a φ-section if $\pi = span\{X, \varphi X\}$ for $X \in \Gamma(TM)$ orthogonal to ξ at each point $p \in M$. The sectional curvature $K(\pi)$ with respect to a φ-section π is called a φ-sectional curvature. If $\{e_1, ..., e_n, \xi\}$ is an orthonormal basis of T_pM and $\{e_{n+1}, ..., e_m\}$ is an orthonormal basis of $T_p^{\perp}M$, then the scalar curvature τ and the normalized scalar curvature ρ at p are defined, respectively, as

$$\tau(p) = \sum_{1 \le i < j \le n} K(e_i \wedge e_j) \qquad \rho = \frac{2\tau}{n(n-1)}.$$

We denote by H the mean curvature vector, that is

$$H(p) = \frac{1}{n} \sum_{i=1}^{n} h(e_i, e_i),$$

and we also set

$$h_{ij}^{\alpha} = g(h(e_i, e_j), e_\alpha), \; i, j \in \{1, ..., n\}, \; \alpha \in \{n+1, ..., m\}.$$

It is well-known that an intrinsic invariant of the submanifold M in $\overline{M}$ is defined by

$$\|H\|^2 = \frac{1}{n^2} \sum_{\alpha=n+1}^{m} \left(\sum_{i=1}^{n} h_{ii}^{\alpha} \right)^2,$$

and the squared norm of h over the dimension n is denoted by $\mathcal{C}$, called the Casorati curvature of the submanifold M. That is,

$$\mathcal{C} = \frac{1}{n} \sum_{\alpha=n+1}^{m} \sum_{i,j=1}^{n} \left(h_{ij}^{\alpha} \right)^2.$$

The submanifold M is said to be *invariantly quasi-umbilical* if there exist $m-n$ mutually orthogonal unit normal vectors $\xi_{n+1}, ..., \xi_m$ such that the shape operator with respect to each direction ξ_α has an eigenvalue of multiplicity $n-1$ and the distinguished eigendirection is the same for each ξ_α.

Suppose now that L is a s-dimensional subspace of T_pM, and $s \geq 2$. Let $\{e_1, ..., e_s\}$ be an orthonormal basis of L. Then the scalar curvature $\tau(L)$ of the s-plane section L is given by

$$\tau(L) = \sum_{1 \leq \alpha < \beta \leq s} K(e_\alpha \wedge e_\beta),$$

and the Casorati curvature $\mathcal{C}(L)$ of the subspace L is defined as

$$\mathcal{C}(L) = \frac{1}{s} \sum_{\alpha=n+1}^{m} \sum_{i,j=1}^{s} \left(h_{ij}^{\alpha}\right)^2.$$

The normalized δ-Casorati curvatures $\delta_c(n-1)$ and $\widehat{\delta}_c(n-1)$ of the submanifold M^n are given by

$$[\delta_c(n-1)]_p = \frac{1}{2}\mathcal{C}_p + \frac{n+1}{2n}\inf\{\mathcal{C}(L) | L \text{ a hyperplane of } T_pM\}$$

and

$$\left[\widehat{\delta}_c(n-1)\right]_p = 2\mathcal{C}_p - \frac{2n-1}{2n}\sup\{\mathcal{C}(L) | L \text{ a hyperplane of } T_pM\}.$$

The generalized normalized δ-Casorati curvatures $\delta_C(t; n-1)$ and $\widehat{\delta}_C(t; n-1)$ of the submanifold M^n are defined for any positive real number $t \neq n(n-1)$ as

$$[\delta_C(t; n-1)]_p = t\mathcal{C}_p + \frac{(n-1)(n+t)(n^2-n-t)}{nt}\inf\{\mathcal{C}(L) | L \text{ a hyperplane of } T_pM\},$$

if $0 < t < n^2 - n$, and

$$\left[\widehat{\delta}_C(t; n-1)\right]_p = t\mathcal{C}_p - \frac{(n-1)(n+t)(t-n^2+n)}{nt}\sup\{\mathcal{C}(L) | L \text{ a hyperplane of } T_pM\}$$

if $t > n^2 - n$.

The C-Bochner curvature tensor [22] on a Sasakian manifold is defined by

$$\begin{aligned} B(X,Y)Z = {} & \overline{R}(X,Y)Z + \frac{1}{2n+4}\{\overline{g}(X,Z)QY - Ric(Y,Z)X \\ & -\overline{g}(Y,Z)QX + Ric(X,Z)Y + \overline{g}(\varphi X,Z)Q\varphi Y \\ & - Ric(\varphi Y,Z)\varphi X - \overline{g}\varphi Y,Z)Q\varphi X + Ric(\varphi X,Z)\varphi Y \\ & + 2Ric(\varphi X,Y)\varphi Z + 2\overline{g}(\varphi X,Y)Q\varphi Z + \eta(Y)\eta(Z)QX \\ & - \eta(Y)Ric(X,Z)\xi + \eta(X)Ric(Y,Z)\xi - \eta(X)\eta(Z)QY\} \\ & - \frac{D+2n}{2n+4}\{\overline{g}(\varphi X,Z)\varphi Y - \overline{g}(\varphi Y,Z)\varphi X + 2\overline{g}(\varphi X,Y)\varphi Z\} \\ & + \frac{D+2n}{2n+4}\{\eta(Y)\overline{g}(X,Z)\xi - \eta(Y)\eta(Z)X + \eta(X)\eta(Z)Y \\ & - \eta(X)\overline{g}(Y,Z)\xi\} - \frac{D-4}{2n+4}\{\overline{g}(X,Z)Y - \overline{g}(Y,Z)X\} \end{aligned} \tag{2}$$

for all $X, Y, Z, W \in \Gamma(T\overline{M})$, where $D = \frac{\tau+2n}{2n+2}$, and $\overline{R}$, Ric, and Q are the Riemannian curvature tensor, the Ricci tensor, and the Ricci operator, respectively. If the C-Bochner curvature tensor vanishes, from Equation (5), we have

$$\begin{aligned}\overline{R}(X,Y,Z,W) = &-\frac{1}{2n+4}\{\overline{g}(X,Z)Ric(Y,W) - Ric(Y,Z)\overline{g}(X,W)\\ &-\overline{g}(Y,Z)Ric(X,W) + Ric(X,Z)\overline{g}(Y,W)\\ &+\overline{g}(\varphi X,Z)Ric(\varphi Y,W) - Ric(\varphi Y,Z)\overline{g}(\varphi X,W)\\ &-\overline{g}(\varphi Y,Z)Ric(\varphi X,W) + Ric(\varphi X,Z)\overline{g}(\varphi Y,W)\\ &+2Ric(\varphi X,Y)\overline{g}(\varphi Z,W) + 2\overline{g}(\varphi X,Y)Ric(\varphi Z,W)\\ &+\eta(Y)\eta(Z)Ric(X,W) - \eta(Y)\eta(W)Ric(X,Z)\\ &+\eta(X)\eta(W)Ric(Y,Z) - \eta(X)\eta(Z)Ric(Y,W)\}\\ &+\frac{D+2n}{2n+4}\{\overline{g}(\varphi X,Z)\overline{g}(\varphi Y,W) - \overline{g}(\varphi Y,Z)\overline{g}(\varphi X,W)\\ &+2\overline{g}(\varphi X,Y)\overline{g}(\varphi Z,W)\} - \frac{D+2n}{2n+4}\{\eta(Y)\eta(W)\overline{g}(X,Z)\\ &-\eta(Y)\eta(Z)\overline{g}(X,W) + \eta(X)\eta(Z)\overline{g}(Y,W)\\ &-\eta(X)\eta(W)\overline{g}(Y,Z)\} + \frac{D-4}{2n+4}\{\overline{g}(X,Z)\overline{g}(Y,W)\\ &-\overline{g}(Y,Z)\overline{g}(X,W)\}\end{aligned} \quad . \tag{3}$$

Now, we recall some definitions from literature on submanifolds.

Definition 1. *Let $(\overline{M}, \varphi, \xi, \eta)$ be an almost contact metric manifolds and M be a submanifold isometrically immersed in $\overline{M}$ tangent to the structure vector field ξ. Then M is said to be invariant (anti-invariant) if $\varphi(T_pM) \subseteq T_pM$ $\left(\varphi(T_pM) \subset T_p^{\perp}M\right)$ for every $p \in M$, where T_pMs denote the tangent space of M at the point p. Moreover, M is called a slant submanifold if for all non-zero vector $U \in T_pM$ at a point p, and the angle of $\theta(U)$ between φU and T_pM is constant (i.e., it does not depend on the choice of $p \in M$ and $U \in \Gamma(T_pM) - <\xi(p)>$).*

Let M^n be an n-dimensional submanifold of a Sasakian manifold $(\overline{M}, \overline{g}, \varphi, \xi, \eta)$. For $X \in \Gamma(TM)$, we can write $\varphi X = PX + QX$, where PX and QX are the tangential and the normal components of φX, respectively. The submanifold is said to be *an anti-invariant (invariant) submanifold* if $P = 0 (Q = 0$, respectively$)$. The squared norm of P at $p \in M$ is defined as

$$||P||^2 = \sum_{i,j=1}^{n} \overline{g}^2(\varphi e_i, e_j)$$

where $\{e_1, \cdots, e_n\}$ is an orthonormal basis of T_pM. The structure vector field ξ can be decomposed as

$$\xi = \xi^{\top} + \xi^{\perp}$$

where $\xi^{\top}$ and $\xi^{\perp}$ are the tangential and the normal components of ξ, respectively.

The following constrained extremum problem plays a key role in the proof of our theorems.

Lemma 1. [23] *Let*

$$\Gamma = \{(x_1, x_2, \cdots, x_n) \in \mathbb{R}^n : x_1 + x_2 + \cdots + x_n = k\}$$

be a hyperplane of $\mathbb{R}^n$, *and* $f : \mathbb{R}^n \longrightarrow \mathbb{R}$ *a quadratic form given by*

$$f(x_1, x_2, \cdots, x_n) = a \sum_{i=1}^{n-1} (x_i)^2 + b\,(x_n)^2 - 2 \sum_{1 \le i < j \le n} x_i x_j, \qquad a > 0,\ b > 0.$$

Then, f has the global extreme at the following point:

$$x_1 = x_2 = \cdots = x_{n-1} = \frac{k}{a+1}, \quad x_n = \frac{k}{b+1} = \frac{k(n-1)}{(a+1)b} = (a-n+2)\frac{k}{a+1}$$

provided that

$$b = \frac{n-1}{a-n+2}$$

by the constrained extremum problem.

3. Inequalities Involving a Vanishing C-Bochner Curvature Tensor

Let M be a submanifold of a Sasakian manifold $(\overline{M}, \overline{g}, \varphi, \xi, \eta)$ with a vanishing C-Bochner curvature tensor. Let $p \in M$ and the set $\{e_1, ..., e_n\}$ and $\{e_{n+1}, ..., e_m\}$ be orthonormal bases of T_pM and $T_p^{\perp}M$, respectively. From Equation (3), we have

$$\begin{aligned}\sum_{i,j=1}^{n} \overline{R}(e_i, e_j, e_j, e_i) = {} & \frac{7n^2 + n - 8 + 2(n-1)||\xi^{\perp}||^2}{4(n+1)(n+2)}\tau \\ & - \frac{3}{n+2}\sum_{i,j=1}^{n} \overline{g}(\varphi e_i, e_j) Ric(e_i, \varphi e_j) \\ & + \frac{n(n-1)(2n+3)}{(n+1)(n+2)}||\xi^{\perp}||^2 - \frac{3n+4}{2(n+1)(n+2)}\end{aligned} \tag{4}$$

Combining Equation (1) and Equation (4), we obtain

$$\begin{aligned}2\tau = n^2||H||^2 - nC + & \frac{7n^2 + n - 8 + 2(n-1)||\xi^{\perp}||^2}{4(n+1)(n+2)}\tau \\ - & \frac{3}{n+2}\sum_{i,j=1}^{n} \overline{g}(\varphi e_i, e_j) Ric(e_i, \varphi e_j) \\ + & \frac{n(n-1)(2n+3)}{(n+1)(n+2)}||\xi^{\perp}||^2 - \frac{3n+4}{2(n+1)(n+2)}\end{aligned} \tag{5}$$

We now consider a quadratic polynomial in the components of the second fundamental form:

$$\begin{aligned}\mathcal{P} = {} & tC + \frac{(n-1)(n+t)(n^2-n-t)}{nt}C(L) - \frac{n^2 + 23n + 24 - 2(n-1)||\xi^{\perp}||^2}{4(n+1)(n+2)}\tau \\ & - \frac{3}{n+2}\sum_{i,j=1}^{n} \overline{g}(\varphi e_i, e_j) Ric(e_i, \varphi e_j) + \frac{n(n-1)(2n+3)}{(n+1)(n+2)}||\xi^{\perp}||^2 - \frac{3n+4}{2(n+1)(n+2)}\end{aligned}$$

where L is a hyperplane of T_pM. Without loss of generality, we may assume that $L = span\{e_1, ..., e_{n-1}\}$. Then we derive

$$\begin{aligned}
\mathcal{P} = & \sum_{\alpha=n+1}^{m} \sum_{i=1}^{n-1} \left[\frac{n^2 + n(t-1) - 2t}{r} (h_{ii}^{\alpha})^2 + \frac{2(n+t)}{n} (h_{in}^{\alpha})^2 \right] \\
& + \sum_{\alpha=n+1}^{m} \left[\frac{2(n+t)(n-1)}{t} \sum_{1=i<j}^{n-1} \left(h_{ij}^{\alpha}\right)^2 - 2 \sum_{1=i<j}^{n} h_{ii}^{\alpha} h_{jj}^{\alpha} + \frac{t}{n} (h_{nn}^{\alpha})^2 \right]. \\
\geq & \sum_{\alpha=n+1}^{m} \left[\sum_{i=1}^{n-1} \frac{n^2 + n(t-1) - 2t}{t} (h_{ii}^{\alpha})^2 - 2 \sum_{1=i<j}^{n} h_{ii}^{\alpha} h_{jj}^{\alpha} + \frac{t}{n} (h_{nn}^{\alpha})^2 \right]
\end{aligned} \tag{6}$$

For $\alpha = n+1, \cdots, m$, we consider the quadratic form $f_\alpha : \mathbb{R}^n \longrightarrow \mathbb{R}$ defined by

$$f_\alpha \left(h_{11}^{\alpha}, \cdots, h_{nn}^{\alpha}\right) = \frac{n^2 + n(t-1) - 2t}{t} \sum_{i=1}^{n-1} (h_{ii}^{\alpha})^2 - 2 \sum_{i<j=1}^{n} h_{ii}^{\alpha} h_{jj}^{\alpha} + \frac{t}{n} (h_{nn}^{\alpha})^2. \tag{7}$$

We then have the constrained extremum problem

$$\min f_\alpha$$

$$\text{subject to } F^\alpha : h_{11}^{\alpha} + \cdots + h_{nn}^{\alpha} = c^\alpha$$

where c^α is a real constant. Comparing Equation (7) with the quadratic function in Lemma 1, we get

$$a = \frac{n^2 + n(t-1) - 2t}{t}, \qquad b = \frac{t}{n}.$$

Therefore, we have the critical point $\left(h_{11}^{\alpha}, \cdots, h_{nn}^{\alpha}\right)$, given by

$$h_{11}^{\alpha} = h_{22}^{\alpha} = \cdots = h_{n-1\,n-1}^{\alpha} = \frac{tc^\alpha}{(n+t)(n-1)}, \qquad h_{nn}^{\alpha} = \frac{nc^\alpha}{n+t},$$

which is a global minimum point by Lemma 1. Moreover, $f_\alpha \left(h_{11}^{\alpha}, \cdots, h_{nn}^{\alpha}\right) = 0$. Therefore, we have

$$\mathcal{P} \geq 0,$$

which implies

$$\begin{aligned}
\frac{n^2 + 23n + 24 - 2(n-1)||\xi^\perp||^2}{4(n+1)(n+2)} \tau \leq tC + & \frac{(n-1)(n+t)(n^2 - n - t)}{nt} C(L) \\
& - \frac{3}{n+2} \sum_{i,j=1}^{n} \overline{g}(\varphi e_i, e_j) Ric(e_i, \varphi e_j) \\
& + \frac{n(n-1)(2n+3)}{(n+1)(n+2)} ||\xi^\perp||^2 - \frac{3n+4}{2(n+1)(n+2)}
\end{aligned}.$$

Therefore, we derive

$$\begin{aligned}\rho \leq & \frac{8(n+1)(n+2)}{n(n-1)\left(n^2+23n+24-2(n-1)||\xi^{\perp}||^2\right)}\left(tC+\frac{(n-1)(n+t)(n^2-n-t)}{nt}C(L)\right) \\ & -\frac{24(n+1)}{n(n-1)\left(n^2+23n+24-2(n-1)||\xi^{\perp}||^2\right)}\sum_{i,j=1}^{n}\overline{g}(\varphi e_i,e_j)Ric(e_i,\varphi e_j) \\ & +\frac{4(2n+3)||\xi^{\perp}||^2}{n^2+23n+24-2(n-1)||\xi^{\perp}||^2}-\frac{4(3n+4)}{n(n-1)\left(n^2+23n+24-2(n-1)||\xi^{\perp}||^2\right)}\end{aligned}.$$

Summing up, we obtain the following theorem:

Theorem 1. *Let M be a submanifold of a Sasakian manifold $(\overline{M},\overline{g},\varphi,\xi,\eta)$ with a vanishing C-Bochner curvature tensor. When $0<t<n^2-n$, the generalized normalized δ-Casorati curvature $\delta_C(t,n-1)$ on M^n satisfies*

$$\begin{aligned}\rho \leq & \frac{8(n+1)(n+2)}{n(n-1)\left(n^2+23n+24-2(n-1)||\xi^{\perp}||^2\right)}\delta_C(t,n-1) \\ & -\frac{24(n+1)}{n(n-1)\left(n^2+23n+24-2(n-1)||\xi^{\perp}||^2\right)}\sum_{i,j=1}^{n}\overline{g}(\varphi e_i,e_j)Ric(e_i,\varphi e_j) \\ & +\frac{4(2n+3)||\xi^{\perp}||^2}{n^2+23n+24-2(n-1)||\xi^{\perp}||^2}-\frac{4(3n+4)}{n(n-1)\left(n^2+23n+24-2(n-1)||\xi^{\perp}||^2\right)}\end{aligned}.$$

Moreover, the equality case holds if and only if M^n is an invariantly quasi-umbilical submanifold with the trivial normal connection in a Sasakian manifold $(\overline{M},\overline{g},\varphi,\xi,\eta)$, such that the shape operators $A_r\equiv A_{\xi_r}$ and $r\in\{n+1,\cdots,m\}$ take the following forms:

$$A_{n+1}=\begin{pmatrix} a & 0 & 0 & \dots & 0 & 0 \\ 0 & a & 0 & \dots & 0 & 0 \\ 0 & 0 & a & \dots & 0 & 0 \\ \vdots & \vdots & \vdots & \ddots & \vdots & \vdots \\ 0 & 0 & 0 & \dots & a & 0 \\ 0 & 0 & 0 & \dots & 0 & \frac{n(n-1)}{t}a \end{pmatrix},\quad A_{n+2}=\cdots=A_m=0 \tag{8}$$

with respect to a suitable orthonormal tangent frame $\{\xi_1,\cdots,\xi_n\}$ and a normal orthonormal frame $\{\xi_{n+1},\cdots,\xi_m\}$.

When a submanifold M is Einstein of a Sasakian manifold $(\overline{M},\overline{g},\varphi,\xi,\eta)$, the Ricci curvature tensor $\rho(X,Y)=\lambda g(X,Y)$ for $X,Y\in\Gamma(TM)$, where λ is some constant. Therefore, we have the following corollary:

Corollary 1. *Let M be an Einstein submanifold of a Sasakian manifold $(\overline{M},\overline{g},\varphi,\xi,\eta)$ with a vanishing C-Bochner curvature tensor. Then, for a Ricci curvature λ, we obtain*

$$\begin{aligned}\rho \leq & \frac{8(n+1)(n+2)}{n(n-1)\left(n^2+23n+24-2(n-1)||\xi^{\perp}||^2\right)}\delta_C(t,n-1) \\ & +\frac{24(n+1)||P||^2\lambda}{n(n-1)\left(n^2+23n+24-2(n-1)||\xi^{\perp}||^2\right)} \\ & +\frac{4(2n+3)||\xi^{\perp}||^2}{n^2+23n+24-2(n-1)||\xi^{\perp}||^2}-\frac{4(3n+4)}{n(n-1)\left(n^2+23n+24-2(n-1)||\xi^{\perp}||^2\right)}\end{aligned}.$$

Moreover, the equality case holds if and only if M^n is an invariantly quasi-umbilical submanifold with the trivial normal connection in a Sasakian manifold $(\overline{M}, \overline{g}, \varphi, \xi, \eta)$, such that with respect to a suitable orthonormal tangent frame $\{\xi_1, \cdots, \xi_n\}$ and a normal orthonormal frame $\{\xi_{n+1}, \cdots, \xi_m\}$, the shape operators $A_r \equiv A_{\xi_r}$ and $r \in \{n+1, \cdots, m\}$ take the form of Equation (8).

For a slant submanifolds ($\overline{g}(\varphi e_i, e_j) = \cos\theta$ with the slant angle θ) of a Sasakian manifold $(\overline{M}, \overline{g}, \varphi, \xi, \eta)$ with a vanishing C-Bochner curvature tensor, we have following corollaries.

Corollary 2. *Let M be a slant submanifold of a Sasakian manifold $(\overline{M}, \overline{g}, \varphi, \xi, \eta)$ with a vanishing C-Bochner curvature tensor. We then obtain*

$$\begin{aligned}\rho \leq & \frac{8(n+1)(n+2)}{n(n-1)\left(n^2+23n+24-2(n-1)||\xi^{\perp}||^2\right)}\delta_C(t, n-1) \\ & + \frac{24(n+1)\cos\theta}{n(n-1)\left(n^2+23n+24-2(n-1)||\xi^{\perp}||^2\right)}\sum_{i,j=1}^{n} Ric(e_i, \varphi e_j) + \frac{4(2n+3)||\xi^{\perp}||^2}{n^2+23n+24-2(n-1)||\xi^{\perp}||^2} \\ & - \frac{4(3n+4)}{n(n-1)\left(n^2+23n+24-2(n-1)||\xi^{\perp}||^2\right)}\end{aligned}$$

where θ is a slant function. Moreover, the equality case holds if and only if, with respect to a suitable frames $\{e_1, ..., e_n\}$ on M and $\{e_{n+1}, ..., e_m\}$ on $T_p^{\perp}M$, $p \in M$, the components of h satisfy

$$\begin{aligned} h_{11}^{\alpha} = h_{22}^{\alpha} = \cdots = h_{n-1\,n-1}^{\alpha} = \tfrac{t}{n(n-1)} h_{nn}^{\alpha}, \qquad & \alpha \in \{n+1, \cdots, m\}, \\ h_{ij}^{\alpha} = 0, \quad i, j \in \{1, 2, \cdots, n\} (i \neq j), \quad & \alpha \in \{n+1, \cdots, m\}. \end{aligned}$$

When the slant angle is zero in Corollary 2, we have the following corollary:

Corollary 3. *Let M be an invariant submanifold of a Sasakian manifold $(\overline{M}, \overline{g}, \varphi, \xi, \eta)$ with a vanishing C-Bochner curvature tensor. We then obtain*

$$\begin{aligned}\rho \leq & \frac{8(n+1)(n+2)}{n(n-1)\left(n^2+23n+24-2(n-1)||\xi^{\perp}||^2\right)}\delta_C(t, n-1) \\ & + \frac{4(6n^2-3n-10)}{n(n-1)\left(n^2+23n+24-2(n-1)||\xi^{\perp}||^2\right)} + \frac{4(2n+3)||\xi^{\perp}||^2}{n^2+23n+24-2(n-1)||\xi^{\perp}||^2}.\end{aligned}$$

Moreover, the equality case holds if and only if, with respect to a suitable frames $\{e_1, ..., e_n\}$ on M and $\{e_{n+1}, ..., e_m\}$ on $T_p^{\perp}M$, $p \in M$, the components of h satisfy

$$\begin{aligned} h_{11}^{\alpha} = h_{22}^{\alpha} = \cdots = h_{n-1\,n-1}^{\alpha} = \tfrac{t}{n(n-1)} h_{nn}^{\alpha}, \qquad & \alpha \in \{n+1, \cdots, m\}, \\ h_{ij}^{\alpha} = 0, \quad i, j \in \{1, 2, \cdots, n\} (i \neq j), \quad & \alpha \in \{n+1, \cdots, m\}. \end{aligned}$$

When the slant angle is $\frac{\pi}{2}$ in Corollary 1, we have the following corollary:

Corollary 4. *Let M be an anti-invariant submanifold of a Sasakian manifold $(\overline{M}, \overline{g}, \varphi, \xi, \eta)$ with a vanishing C-Bochner curvature tensor. We then obtain*

$$\begin{aligned}\rho \leq & \frac{8(n+1)(n+2)}{n(n-1)\left(n^2+23n+24-2(n-1)||\xi^{\perp}||^2\right)}\delta_C(t, n-1) \\ & + \frac{4(2n+3)||\xi^{\perp}||^2}{n^2+23n+24-2(n-1)||\xi^{\perp}||^2} - \frac{4(3n+4)}{n(n-1)\left(n^2+23n+24-2(n-1)||\xi^{\perp}||^2\right)}.\end{aligned}$$

Moreover, the equality case holds if and only if, with respect to a suitable frames $\{e_1, ..., e_n\}$ on M and $\{e_{n+1}, ..., e_m\}$ on $T_p^{\perp}M$, $p \in M$, the components of h satisfy

$$h_{11}^{\alpha} = h_{22}^{\alpha} = \cdots = h_{n-1\,n-1}^{\alpha} = \frac{t}{n(n-1)} h_{nn}^{\alpha}, \qquad \alpha \in \{n+1, \cdots, m\},$$
$$h_{ij}^{\alpha} = 0, \quad i,j \in \{1,2,\cdots,n\} (i \neq j), \quad \alpha \in \{n+1, \cdots, m\}.$$

Remark 1. *In the case for $t > n^2 - n$, the methods of finding the above inequailities is analogous. Thus, we leave these problems for readers.*

Taking $t = \frac{n(n-1)}{2}$ in $\delta_C(t, n-1)$, we have the following relation:

$$\left[\delta_C\left(\frac{n(n-1)}{2}; n-1\right)\right]_p = n(n-1)\left[\delta_C(n-1)\right]_p$$

in any point $p \in M$. Therefore, we have following optimal inequalities for the normalized δ-Casorati curvature $\delta_C(n-1)$.

Corollary 5. *Let M be a submanifold of a Sasakian manifold $(\overline{M}, \overline{g}, \varphi, \xi, \eta)$ with a vanishing C-Bochner curvature tensor. The normalized δ-Casorati curvature $\delta_C(n-1)$ on M^n satisfies*

$$\begin{aligned}\rho \leq\ & \frac{8(n+1)(n+2)}{(n^2+23n+24-2(n-1)||\xi^{\perp}||^2)}\delta_C(n-1) \\ & - \frac{24(n+1)}{n(n-1)\left(n^2+23n+24-2(n-1)||\xi^{\perp}||^2\right)}\sum_{i,j=1}^{n}\overline{g}(\varphi e_i, e_j) Ric(e_i, \varphi e_j) \\ & + \frac{4(2n+3)||\xi^{\perp}||^2}{n^2+23n+24-2(n-1)||\xi^{\perp}||^2} - \frac{4(3n+4)}{n(n-1)\left(n^2+23n+24-2(n-1)||\xi^{\perp}||^2\right)}\end{aligned}.$$

Corollary 6. *Let M be an Einstein submanifold of a Sasakian manifold $(\overline{M}, \overline{g}, \varphi, \xi, \eta)$ with a vanishing C-Bochner curvature tensor. Then, for a Ricci curvature λ, we obtain*

$$\begin{aligned}\rho \leq\ & \frac{8(n+1)(n+2)}{(n^2+23n+24-2(n-1)||\xi^{\perp}||^2)}\delta_C(n-1) \\ & + \frac{24(n+1)||P||^2\lambda}{n(n-1)\left(n^2+23n+24-2(n-1)||\xi^{\perp}||^2\right)} + \frac{4(2n+3)||\xi^{\perp}||^2}{n^2+23n+24-2(n-1)||\xi^{\perp}||^2} \\ & - \frac{4(3n+4)}{n(n-1)\left(n^2+23n+24-2(n-1)||\xi^{\perp}||^2\right)}\end{aligned}.$$

Corollary 7. *Let M be a slant submanifold of a Sasakian manifold $(\overline{M}, \overline{g}, \varphi, \xi, \eta)$ with a vanishing C-Bochner curvature tensor. We then obtain*

$$\begin{aligned}\rho \leq\ & \frac{8(n+1)(n+2)}{(n^2+23n+24-2(n-1)||\xi^{\perp}||^2)}\delta_C(n-1) \\ & + \frac{24(n+1)\cos\theta}{n(n-1)\left(n^2+23n+24-2(n-1)||\xi^{\perp}||^2\right)}\sum_{i,j=1}^{n} Ric(e_i, \varphi e_j) \\ & + \frac{4(2n+3)||\xi^{\perp}||^2}{n^2+23n+24-2(n-1)||\xi^{\perp}||^2} - \frac{4(3n+4)}{n(n-1)\left(n^2+23n+24-2(n-1)||\xi^{\perp}||^2\right)}\end{aligned}$$

where θ is a slant function.

Corollary 8. *Let M be an invariant submanifold of a Sasakian manifold* $(\overline{M}, \overline{g}, \varphi, \xi, \eta)$ *with a vanishing C-Bochner curvature tensor. We then obtain*

$$\rho \leq \frac{8(n+1)(n+2)}{(n^2+23n+24-2(n-1)||\xi^{\perp}||^2)}\delta_C(n-1) + \frac{4(6n^2-3n-10)}{n(n-1)\left(n^2+23n+24-2(n-1)||\xi^{\perp}||^2\right)} + \frac{4(2n+3)||\xi^{\perp}||^2}{n^2+23n+24-2(n-1)||\xi^{\perp}||^2}.$$

Corollary 9. *Let M be an anti-invariant submanifold of a Sasakian manifold* $(\overline{M}, \overline{g}, \varphi, \xi, \eta)$ *with a vanishing C-Bochner curvature tensor. We then obtain*

$$\rho \leq \frac{8(n+1)(n+2)}{(n^2+23n+24-2(n-1)||\xi^{\perp}||^2)}\delta_C(n-1) + \frac{4(2n+3)||\xi^{\perp}||^2}{n^2+23n+24-2(n-1)||\xi^{\perp}||^2} - \frac{4(3n+4)}{n(n-1)\left(n^2+23n+24-2(n-1)||\xi^{\perp}||^2\right)}.$$

Author Contributions: C.W.L. gave the idea to establish optimizations to a vanishing Bochner curvature tensor. J.W.L. checked and polished the draft.

Funding: Jae Won Lee was supported by Basic Science Research Program through the National Research Foundation of Korea (NRF) funded by the Ministry of Education (2017R1D1A1B03033978) and Chul Woo Lee was supported by Basic Science Research Program through the National Research Foundation of Korea (NRF) funded by the Ministry of Education (2018R1D1A1B07040576).

Acknowledgments: The authors thank the reviewers for their valuable suggestions to improve the quality and presentation of this paper.

Conflicts of Interest: The authors declare no conflict of interest.

References

1. Bochner, S. Curvature and Betti numbers II. *Ann. Math.* **1949**, *50*, 77–93. [CrossRef]
2. Webster, S.M. On the pseudo-conformal geometry of a Kähler manifold. *Math. Z.* **1977**, *157*, 265–270. [CrossRef]
3. Matsumoto, M.; Chuman, G. On C-Bochner curvature tensor. *TRU Math.* **1969**, *5*, 21–30.
4. Kim, B.H. Fibred Sasakian spaces with vanishing contact Bochner curvature tensor. *Hiroshima Math. J.* **1989**, *19*, 181–195.
5. Tano, S. The topology of contact Riemannian manifolds. *Ill. J. Math.* **1968**, *12*, 700–712.
6. Casorati, F. Mesure de la courbure des surfaces suivant l'idée commune. Ses rapports avec les mesures de courbure gaussienne et moyenne. *Acta Math.* **1890**, *14*, 95–110. [CrossRef]
7. Kowalczyk, D. Casorati curvatures. *Bull. Trans. Univ. Brasov. Ser. III* **2008**, *1*, 209–213.
8. Decu, S.; Haesen, S.; Verstraelen, L. Optimal inequalities involving Casorati curvatures. *Bull. Transilv. Univ. Braşov Ser. B* **2007**, *14*, 85–93.
9. Decu, S.; Haesen, S.; Verstraelen, L. Optimal inequalities characterising quasi-umbilical submanifolds. *J. Inequal. Pure Appl. Math.* **2008**, *9*, 1–7.
10. Ghişoiu, V. Inequalities for the Casorati curvatures of slant submanifolds in complex space forms. In Proceedings of the RIGA Riemannian Geometry and Applications, Bucharest, Romania, 10–14 May 2011; pp. 145–150.
11. He, G.; Liu, H.; Zhang, L. Optimal inequalities for the Casorati curvatures of submanifolds in generalized space forms endowed with semi-symmetric non-metric connections. *Symmetry* **2016**, *8*, 10. [CrossRef]
12. Lee, C.W.; Yoon, D.W.; Lee, J.W. Optimal inequalities for the Casorati curvatures of submanifolds of real space forms endowed with semi-symmetric metric connections. *J. Inequal. Appl.* **2014**, *2014*, 327. [CrossRef]
13. Lee, C.W.; Lee, J.W.; Vîlcu, G.-E.; Yoon, D.W. Optimal inequalities for the Casorati curvatures of submanifolds of generalized space forms endowed with semi-symmetric metric connections. *Bull. Korean Math. Soc.* **2015**, *52*, 1631–1647. [CrossRef]
14. Lee, C.W.; Lee, J.W.; Vîlcu, G.-E. A new proof for some optimal inequalities involving generalized normalized δ-Casorati curvatures. *J. Inequal. Appl.* **2015**, *2015*, 30. [CrossRef]

15. Lee, J.W.; Lee, C.W.; Yoon, D.W. Inequalities for generalized δ-Casorati curvatures of submanifolds in real space forms endowed with a semi-symmetric metric connection. *Rev. Union Mat. Argent.* **2016**, *57*, 53–62.
16. Lee, J.W.; Vîlcu, G.-E. Inequalities for generalized normalized δ-Casorati curvatures of slant submanifolds in quaternionic space forms. *Taiwan J. Math.* **2015**, *19*, 691–702. [CrossRef]
17. Slesar, V.; Şahin, B.; Vîlcu, G.-E. Inequalities for the Casorati curvatures of slant submanifolds in quaternionic space forms. *J. Inequal. Appl.* **2014**, *2014*, 123. [CrossRef]
18. Zhang, P.; Zhang, L. Casorati inequalities for submanifolds in a Riemannian manifold of quasi-constant curvature with a semi-symmetric metric connection. *Symmetry* **2016**, *8*, 19. [CrossRef]
19. Zhang, P.; Zhang, L. Inequalities for Casorati curvatures of submanifolds in real space forms. *Adv. Geom.* **2016**, *16*, 329–335. [CrossRef]
20. Blair, D.E. Contact manifold in Riemannian geometry. In *Lecture Notes in Math*; Springer-Verlag: Berlin/Heidelberg, Germany, 1976.
21. Sasaki, S.; Hatakeyama, Y. On differentiable manifolds with contact metric structure. *J. Math. Soc. Jpn.* **1961**, *14*, 249–271. [CrossRef]
22. Kim, J.S.; Tripathi, M.M.; Choi, J. On C-Bochner curvature tensor of a contact metric manifold. *Bull. Korean Math. Soc.* **2005**, *42*, 713–724. [CrossRef]
23. Tripathi, M.M. Inequalities for algebraic Casorati curvatures and their applications. *Note Mat.* **2017**, *37*, 161–186.

mathematics

MDPI

Article

Curvature Invariants for Statistical Submanifolds of Hessian Manifolds of Constant Hessian Curvature

Adela Mihai [1] **and Ion Mihai** [2,*]

1 Department of Mathematics and Computer Science, Technical University of Civil Engineering Bucharest, 020396 Bucharest, Romania; adela.mihai@utcb.com
2 Department of Mathematics, Faculty of Mathematics and Computer Science, University of Bucharest, 010014 Bucharest, Romania
* Correspondence: imihai@fmi.unibuc.ro

Received: 23 February 2018; Accepted: 12 March 2018; Published: 15 March 2018

Abstract: We consider statistical submanifolds of Hessian manifolds of constant Hessian curvature. For such submanifolds we establish a Euler inequality and a Chen-Ricci inequality with respect to a sectional curvature of the ambient Hessian manifold.

Keywords: statistical manifolds; Hessian manifolds; Hessian sectional curvature; scalar curvature; Ricci curvature

MSC: Math. Subject Classification: 53C05, 53C40.

1. Introduction

It is well-known that curvature invariants play the most fundamental role in Riemannian geometry. Curvature invariants provide the intrinsic characteristics of Riemannian manifolds which affect the behavior in general of the Riemannian manifold. They are the main Riemannian invariants and the most natural ones. Curvature invariants also play key roles in physics. For instance, the magnitude of a force required to move an object at constant speed, according to Newton's laws, is a constant multiple of the curvature of the trajectory. The motion of a body in a gravitational field is determined, according to Einstein's general theory of relativity, by the curvatures of spacetime. All sorts of shapes, from soap bubbles to red cells are determined by various curvatures.

Classically, among the curvature invariants, the most studied were sectional, scalar and Ricci curvatures.

Chen [1] established a generalized Euler inequality for submanifolds in real space forms. Also a sharp relationship between the Ricci curvature and the squared mean curvature for any Riemannian submanifold of a real space form was proved in [2], which is known as the Chen-Ricci inequality.

Statistical manifolds introduced, in 1985, by Amari have been studied in terms of information geometry. Since the geometry of such manifolds includes the notion of dual connections, also called conjugate connections in affine geometry, it is closely related to affine differential geometry. Further, a statistical structure is a generalization of a Hessian one.

In [3], Aydin and the present authors obtained geometrical inequalities for the scalar curvature and the Ricci curvature associated to the dual connections for submanifolds in statistical manifolds of constant curvature. We want to point-out that, generally, the dual connections are not metric; then one cannot define a sectional curvature with respect to them by the standard definitions. However there exists a sectional curvature on a statistical manifold defined by B. Opozda (see [4]).

We mention that in [5] we established a Wintgen inequality for statistical submanifolds in statistical manifolds of constant curvature by using another sectional curvature.

As we know, submanifolds in Hessian manifolds have not been considered until now.

In the present paper we deal with statistical submanifolds in Hessian manifolds of constant Hessian curvature c. It is known [6] that such a manifold is a statistical manifold of null constant curvature and also a Riemannian space form of constant sectional curvature $-c/4$ (with respect to the sectional curvature defined by the Levi-Civita connection).

2. Statistical Manifolds and Their Submanifolds

A *statistical manifold* is an m-dimensional Riemannian manifold $(\tilde{M}^m, g)$ endowed with a pair of torsion-free affine connections $\tilde{\nabla}$ and $\tilde{\nabla}^*$ satisfying

$$Z\tilde{g}(X,Y) = \tilde{g}(\tilde{\nabla}_Z X, Y) + \tilde{g}(X, \tilde{\nabla}^*_Z Y), \tag{1}$$

for any $X, Y, Z \in \Gamma(T\tilde{M}^m)$. The connections $\tilde{\nabla}$ and $\tilde{\nabla}^*$ are called *dual connections* (see [7–9]), and it is easily shown that $(\tilde{\nabla}^*)^* = \tilde{\nabla}$. The pair $(\tilde{\nabla}, g)$ is said to be a *statistical structure*. If $(\tilde{\nabla}, g)$ is a statistical structure on $\tilde{M}^m$, then $(\tilde{\nabla}^*, g)$ is a statistical structure too [8].

On the other hand, any torsion-free affine connection $\tilde{\nabla}$ always has a dual connection given by

$$\tilde{\nabla} + \tilde{\nabla}^* = 2\tilde{\nabla}^0, \tag{2}$$

where $\tilde{\nabla}^0$ is the Levi-Civita connection on $\tilde{M}^m$.

Denote by $\tilde{R}$ and $\tilde{R}^*$ the curvature tensor fields of $\tilde{\nabla}$ and $\tilde{\nabla}^*$, respectively.

A statistical structure $(\tilde{\nabla}, g)$ is said to be *of constant curvature* $\varepsilon \in \mathbb{R}$ if

$$\tilde{R}(X,Y)Z = \varepsilon\{g(Y,Z)X - g(X,Z)Y\}. \tag{3}$$

A statistical structure $(\tilde{\nabla}, g)$ of constant curvature 0 is called a *Hessian structure*.

The curvature tensor fields $\tilde{R}$ and $\tilde{R}^*$ of the dual connections satisfy

$$g(\tilde{R}^*(X,Y)Z, W) = -g(Z, \tilde{R}(X,Y)W). \tag{4}$$

From (4) it follows immediately that if $(\tilde{\nabla}, g)$ is a statistical structure of constant curvature ε, then $(\tilde{\nabla}^*, g)$ is also a statistical structure of constant curvature ε. In particular, if $(\tilde{\nabla}, g)$ is Hessian, $(\tilde{\nabla}^*, g)$ is also Hessian [6].

On a Hessian manifold $(\tilde{M}^m, \tilde{\nabla})$, let $\gamma = \tilde{\nabla} - \tilde{\nabla}^0$. The tensor field $\tilde{Q}$ of type (1, 3) defined by $\tilde{Q}(X,Y) = [\gamma_X, \gamma_Y]$, $X, Y \in \Gamma(T\tilde{M}^m)$ is said to be the *Hessian curvature tensor* for $\tilde{\nabla}$ (see [4,6]). It satisfies

$$\tilde{R}(X,Y) + \tilde{R}^*(X,Y) = 2\tilde{R}^0(X,Y) + 2\tilde{Q}(X,Y).$$

By using the Hessian curvature tensor $\tilde{Q}$, a Hessian sectional curvature can be defined on a Hessian manifold.

Let $p \in \tilde{M}^m$ and π a plane in $T_p\tilde{M}^m$. Take an orthonormal basis $\{X, Y\}$ of π and set

$$\tilde{K}(\pi) = g(\tilde{Q}(X,Y)Y, X).$$

The number $\tilde{K}(\pi)$ is independent of the choice of an orthonormal basis and is called the *Hessian sectional curvature*.

A Hessian manifold has constant Hessian sectional curvature c if and only if (see [6])

$$\tilde{Q}(X,Y,Z,W) = \frac{c}{2}[g(X,Y)g(Z,W) + g(X,W)g(Y,Z)],$$

for all vector fields on $\tilde{M}^m$.

If $(\tilde{M}^m, g)$ is a statistical manifold and M^n an n-dimensional submanifold of $\tilde{M}^m$, then (M^n, g) is also a statistical manifold with the induced connection by $\tilde{\nabla}$ and induced metric g. In the case that

$(\tilde{M}^m, g)$ is a semi-Riemannian manifold, the induced metric g has to be non-degenerate. For details, see [10].

In the geometry of Riemannian submanifolds (see [11]), the fundamental equations are the Gauss and Weingarten formulas and the equations of Gauss, Codazzi and Ricci.

Let denote the set of the sections of the normal bundle to M^n by $\Gamma(T^{\perp}M^n)$.

In our case, for any $X, Y \in \Gamma(TM^n)$, according to [10], the corresponding Gauss formulas are

$$\tilde{\nabla}_X Y = \nabla_X Y + h(X, Y), \tag{5}$$

$$\tilde{\nabla}^*_X Y = \nabla^*_X Y + h^*(X, Y), \tag{6}$$

where $h, h^* : \Gamma(TM^n) \times \Gamma(TM^n) \to \Gamma(T^{\perp}M^n)$ are symmetric and bilinear, called the *imbedding curvature tensor* of M^n in $\tilde{M}^m$ for $\tilde{\nabla}$ and the *imbedding curvature tensor* of M^n in $\tilde{M}^m$ for $\tilde{\nabla}^*$, respectively.

In [10], it is also proved that (∇, g) and (∇^*, g) are dual statistical structures on M^n.

Since h and h^* are bilinear, there exist linear transformations A_ξ and A^*_ξ on TM^n defined by

$$g(A_\xi X, Y) = g(h(X, Y), \xi), \tag{7}$$

$$g(A^*_\xi X, Y) = g(h^*(X, Y), \xi), \tag{8}$$

for any $\xi \in \Gamma(T^{\perp}M^n)$ and $X, Y \in \Gamma(TM^n)$. Further, see [10], the corresponding Weingarten formulas are

$$\tilde{\nabla}_X \xi = -A^*_\xi X + \nabla^{\perp}_X \xi, \tag{9}$$

$$\tilde{\nabla}^*_X \xi = -A_\xi X + \nabla^{*\perp}_X \xi, \tag{10}$$

for any $\xi \in \Gamma(T^{\perp}M^n)$ and $X \in \Gamma(TM^n)$. The connections $\nabla^{\perp}_X$ and $\nabla^{*\perp}_X$ given by (9) and (10) are Riemannian dual connections with respect to induced metric on $\Gamma(T^{\perp}M^n)$.

Let $\{e_1, ..., e_n\}$ and $\{e_{n+1}, ..., e_m\}$ be orthonormal tangent and normal frames, respectively, on M^n. Then the mean curvature vector fields are defined by

$$H = \tfrac{1}{n}\sum_{i=1}^{n} h(e_i, e_i) = \tfrac{1}{n}\sum_{\alpha=n+1}^{m}\left(\sum_{i=1}^{n} h^{\alpha}_{ii}\right)e_\alpha,\ h^{\alpha}_{ij} = g(h(e_i, e_j), e_\alpha), \tag{11}$$

and

$$H^* = \tfrac{1}{n}\sum_{i=1}^{n} h^*(e_i, e_i) = \tfrac{1}{n}\sum_{\alpha=n+1}^{m}\left(\sum_{i=1}^{n} h^{*\alpha}_{ii}\right)e_\alpha,\ h^{*\alpha}_{ij} = g(h^*(e_i, e_j), e_\alpha), \tag{12}$$

for $1 \leq i, j \leq n$ and $n + 1 \leq \alpha \leq m$.

The corresponding Gauss, Codazzi and Ricci equations are given by the following result.

Proposition 1. *[10] Let $\tilde{\nabla}$ and $\tilde{\nabla}^*$ be dual connections on a statistical manifold $\tilde{M}^m$ and ∇ the induced connection by $\tilde{\nabla}$ on a statistical submanifold M^n. Let $\tilde{R}$ and R be the Riemannian curvature tensors for $\tilde{\nabla}$ and ∇, respectively. Then*

$$g(\tilde{R}(X, Y)Z, W) = g(R(X, Y)Z, W) + g(h(X, Z), h^*(Y, W)) - g(h^*(X, W), h(Y, Z)), \tag{13}$$

$$\begin{aligned}(\tilde{R}(X, Y)Z)^{\perp} = {} & \nabla^{\perp}_X h(Y, Z) - h(\nabla_X Y, Z) - h(Y, \nabla_X Z) - \\ & -\{\nabla^{\perp}_Y h(X, Z) - h(\nabla_Y X, Z) - h(X, \nabla_Y Z)\},\end{aligned}$$

$$g(R^{\perp}(X, Y)\xi, \eta) = g(\tilde{R}(X, Y)\xi, \eta) + g([A^*_\xi, A_\eta]X, Y), \tag{14}$$

where $R^{\perp}$ is the Riemannian curvature tensor of $\nabla^{\perp}$ on $T^{\perp}M^n$, $\xi,\eta \in \Gamma(T^{\perp}M^n)$ and $[A^*_{\xi}, A_{\eta}] = A^*_{\xi}A_{\eta} - A_{\eta}A^*_{\xi}$.

For the equations of Gauss, Codazzi and Ricci with respect to the connection $\tilde{\nabla}^*$ on M^n, we have

Proposition 2. *[10] Let $\tilde{\nabla}$ and $\tilde{\nabla}^*$ be dual connections on a statistical manifold $\tilde{M}^m$ and ∇^* the induced connection by $\tilde{\nabla}^*$ on a statistical submanifold M^n. Let $\tilde{R}^*$ and R^* be the Riemannian curvature tensors for $\tilde{\nabla}^*$ and ∇^*, respectively. Then*

$$g(\tilde{R}^*(X,Y)Z,W) = g(R^*(X,Y)Z,W) + g(h^*(X,Z),h(Y,W)) - g(h(X,W),h^*(Y,Z)), \tag{15}$$

$$(\tilde{R}^*(X,Y)Z)^{\perp} = \nabla^{*\perp}_X h^*(Y,Z) - h^*(\nabla^*_X Y,Z) - h^*(Y,\nabla^*_X Z) - $$
$$-\{\nabla^{*\perp}_Y h^*(X,Z) - h^*(\nabla^*_Y X,Z) - h^*(X,\nabla^*_Y Z)\},$$

$$g(R^{*\perp}(X,Y)\xi,\eta) = g(\tilde{R}^*(X,Y)\xi,\eta) + g([A_{\xi},A^*_{\eta}]X,Y), \tag{16}$$

where $R^{*\perp}$ is the Riemannian curvature tensor of $\nabla^{\perp *}$ on $T^{\perp}M^n$, $\xi,\eta \in \Gamma(T^{\perp}M^n)$ and $\left[A_{\xi},A^*_{\eta}\right] = A_{\xi}A^*_{\eta} - A^*_{\eta}A_{\xi}$.

Geometric inequalities for statistical submanifolds in statistical manifolds with constant curvature were obtained in [3].

3. Euler Inequality and Chen-Ricci Inequality

First we obtain a Euler inequality for submanifolds in a Hessian manifold of constant Hessian curvature.

Let $\tilde{M}^m(c)$ be a Hessian manifold of constant Hessian curvature c. Then it is flat with respect to the dual connections $\tilde{\nabla}$ and $\tilde{\nabla}^*$. Moreover $\tilde{M}^m(c)$ is a Riemannian space form of constant sectional curvature $-c/4$ (with respect to the Levi-Civita connection $\tilde{\nabla}^0$).

Let M^n be an n-dimensional statistical submanifold of $\tilde{M}^m(c)$ and $\{e_1,...,e_n\}$ and $\{e_{n+1},...,e_m\}$ be orthonormal tangent and normal frames, respectively, on M^n.

We denote by τ^0 the scalar curvature of the Levi-Civita connection ∇^0 on M^n. Gauss equation implies

$$2\tau^0 = n^2\|H^0\|^2 - \|h^0\|^2 - n(n-1)\tfrac{c}{4}, \tag{17}$$

where H^0 and h^0 are the mean curvature vector and the second fundamental form, respectively, with respect to the Levi-Civita connection.

Let τ be the scalar curvature of M^n (with respect to the Hessian curvature tensor Q). Then, from (13) and (15), we have:

$$2\tau = \frac{1}{2}\sum_{i,j=1}^{n}[g(R(e_i,e_j)e_j,e_i) + g(R^*(e_i,e_j)e_j,e_i) - 2g(R^0(e_i,e_j)e_j,e_i)] =$$

$$= n^2 g(H,H^*) - \sum_{i,j=1}^{n} g(h(e_i,e_j),h^*(e_i,e_j)) - 2\tau^0 =$$

$$= n^2 g(H,H^*) - \sum_{r=n+1}^{m}\sum_{i,j=1}^{n} h^r_{ij}h^{*r}_{ij} - 2\tau^0 =$$

$$= n^2 g(H,H^*) - \frac{1}{4}\sum_{r=n+1}^{m}\sum_{i,j=1}^{n}[(h^r_{ij}+h^{*r}_{ij})^2 - (h^r_{ij}-h^{*r}_{ij})^2] - 2\tau^0 =$$

$$= n^2 g(H,H^*) - \|h^0\|^2 + \frac{1}{4}\sum_{r=n+1}^{m}\sum_{i,j=1}^{n}(h_{ij}^r - h_{ij}^{*r})^2 - 2\tau^0.$$

By (17), it follows that

$$2\tau \geq n^2 g(H,H^*) - n^2\|H^0\|^2 + n(n-1)\frac{c}{4} =$$

$$= n^2 g(H,H^*) - \frac{n^2}{4} g(H+H^*, H+H^*) + n(n-1)\frac{c}{4} =$$

$$= \frac{n^2}{2} g(H,H^*) - \frac{n^2}{4} g(H,H) - \frac{n^2}{4} g(H^*,H^*) + n(n-1)\frac{c}{4} =$$

$$= -\frac{n^2}{4}\|H - H^*\|^2 + n(n-1)\frac{c}{4}.$$

Summing up, we proved the following.

Theorem 1. *Let M^n be a statistical submanifold of a Hessian manifold $\tilde{M}^m(c)$ of constant Hessian curvature c. Then the scalar curvature satisfies:*

$$2\tau \geq -\frac{n^2}{4}\|H - H^*\|^2 + n(n-1)\frac{c}{4}.$$

Moreover, the equality holds at any pont $p \in M^n$ if and only if $h = h^*$. In this case, the scalar curvature is constant, $2\tau = n(n-1)\frac{c}{4}$.

We want to point-out that τ is non-positive on standard examples of Hessian manifolds.

Next we establish a Chen-Ricci inequality for statistical submanifolds in Hessian manifolds of constant Hessian curvature.

Recall that

$$2\tau = \frac{1}{2}\sum_{i,j=1}^{n}[g(R(e_i,e_j)e_j,e_i) + g(R^*(e_i,e_j)e_j,e_i) - 2g(R^0(e_i,e_j)e_j,e_i)] =$$

$$= n^2 g(H,H^*) - \sum_{i,j=1}^{n} g(h(e_i,e_j), h^*(e_i,e_j)) - 2\tau^0.$$

Then

$$2\tau = \frac{n^2}{2}[g(H+H^*, H+H^*) - g(H,H) - g(H^*,H^*)] -$$

$$-\frac{1}{2}[g(h(e_i,e_j) + h^*(e_i,e_j), h(e_i,e_j) + h^*(e_i,e_j)) -$$

$$-g(h(e_i,e_j), h(e_i,e_j)) - g(h^*(e_i,e_j), h^*(e_i,e_j))] - 2\tau^0 =$$

$$= 2n^2 g(H^0,H^0) - 2\left\|h^0\right\|^2 - 2\tau^0 -$$

$$-\frac{n^2}{2} g(H,H) - \frac{n^2}{2} g(H^*,H^*) + \frac{1}{2}\|h\|^2 + \frac{1}{2}\|h^*\|^2,$$

where H^0 and h^0 are the mean curvature vector and the second fundamental form, respectively, with respect to the Levi-Civita connection.

By using (17), we get

$$2\tau = n(n-1)\frac{c}{2} + 2\tau^0 - \frac{n^2}{2} g(H,H) - \frac{n^2}{2} g(H^*,H^*) + \frac{1}{2}\|h\|^2 + \frac{1}{2}\|h^*\|^2. \quad (18)$$

On the other hand, we may write:

$$\|h\|^2 = \sum_{\alpha=n+1}^{m} [(h_{11}^\alpha)^2 + (h_{22}^\alpha + ... + h_{nn}^\alpha)^2 + 2\sum_{1\le i<j\le n}(h_{ij}^\alpha)^2 - 2\sum_{2\le i\ne j\le n} h_{ii}^\alpha h_{jj}^\alpha] =$$

$$= \frac{1}{2}\sum_{\alpha=n+1}^{m} [(h_{11}^\alpha + h_{22}^\alpha + ... + h_{nn}^\alpha)^2 + (h_{11}^\alpha - h_{22}^\alpha - ... - h_{nn}^\alpha)^2] -$$

$$- \sum_{\alpha=n+1}^{m}\sum_{2\le i\ne j\le n} [h_{ii}^\alpha h_{jj}^\alpha - (h_{ij}^\alpha)^2] + 2\sum_{\alpha=n+1}^{m}\sum_{j=1}^{n}(h_{1j}^\alpha)^2 \ge$$

$$\ge \frac{n^2}{2}\|H\|^2 - \sum_{\alpha=n+1}^{m}\sum_{2\le i\ne j\le n} [h_{ii}^\alpha h_{jj}^\alpha - (h_{ij}^\alpha)^2].$$

In the same manner, one obtains

$$\|h^*\|^2 \ge \frac{n^2}{2}\|H^*\|^2 - \sum_{\alpha=n+1}^{m}\sum_{2\le i\ne j\le n} [h_{ii}^{*\alpha} h_{jj}^{*\alpha} - (h_{ij}^{*\alpha})^2].$$

Substituting the above inequalities in (18), it follows that

$$2\tau \ge n(n-1)\frac{c}{2} + 2\tau^0 - \frac{n^2}{4}\|H\|^2 - \frac{n^2}{4}\|H^*\|^2 -$$

$$-\frac{1}{2}\sum_{\alpha=n+1}^{m}\sum_{2\le i\ne j\le n} [h_{ii}^\alpha h_{jj}^\alpha - (h_{ij}^\alpha)^2] - \frac{1}{2}\sum_{\alpha=n+1}^{m}\sum_{2\le i\ne j\le n} [h_{ii}^{*\alpha} h_{jj}^{*\alpha} - (h_{ij}^{*\alpha})^2] =$$

$$= n(n-1)\frac{c}{2} + 2\tau^0 - \frac{n^2}{4}\|H\|^2 - \frac{n^2}{4}\|H^*\|^2 -$$

$$-\frac{1}{2}\sum_{\alpha=n+1}^{m}\sum_{2\le i\ne j\le n} [(h_{ii}^\alpha + h_{ii}^{*\alpha})(h_{jj}^\alpha + h_{jj}^{*\alpha}) - (h_{ij}^\alpha + h_{ij}^{*\alpha})^2] +$$

$$+\frac{1}{2}\sum_{\alpha=n+1}^{m}\sum_{2\le i\ne j\le n} (h_{ii}^\alpha h_{jj}^{*\alpha} - h_{ij}^\alpha h_{ij}^{*\alpha}) + \frac{1}{2}\sum_{\alpha=n+1}^{m}\sum_{2\le i\ne j\le n} (h_{ii}^{*\alpha} h_{jj}^\alpha - h_{ij}^\alpha h_{ij}^{*\alpha}) =$$

$$= n(n-1)\frac{c}{2} + 2\tau^0 - \frac{n^2}{4}\|H\|^2 - \frac{n^2}{4}\|H^*\|^2 -$$

$$-2\sum_{\alpha=n+1}^{m}\sum_{2\le i\ne j\le n} [h_{ii}^{0\alpha} h_{jj}^{0\alpha} - (h_{ij}^{0\alpha})^2] + \frac{1}{2}\sum_{2\le i\ne j\le n} [g(R(e_i,e_j)e_j,e_i) + g(R^*(e_i,e_j)e_j,e_i)].$$

Gauss equation for the Levi-Civita connection and the definition of the Hessian sectional curvature imply

$$2\tau \ge n(n-1)\frac{c}{2} + 2\tau^0 - \frac{n^2}{4}\|H\|^2 - \frac{n^2}{4}\|H^*\|^2 -$$

$$-(n-1)(n-2)\frac{c}{2} - 2\sum_{2\le i\ne j\le n} K^0(e_i\wedge e_j) + \sum_{2\le i\ne j\le n} [K(e_i\wedge e_j) + K^0(e_i\wedge e_j)].$$

But the Ricci curvature R^0 with respect to the Levi-Civita connection is given by

$$2Ric^0(X) = 2\tau^0 - \sum_{2\le i\ne j\le n} K^0(e_i\wedge e_j),$$

and, similarly,

$$2Ric(X) = 2\tau - \sum_{2\leq i\neq j\leq n} K(e_i \wedge e_j).$$

Consequently

$$Ric(X) \geq (n-1)\frac{c}{2} - \frac{n^2}{8}\|H\|^2 - \frac{n^2}{8}\|H^*\|^2 + Ric^0(X).$$

The vector field $X = e_1$ satisfies the equality case if and only if

$$\begin{cases} h_{11}^{\alpha} = h_{22}^{\alpha} + ... + h_{nn}^{\alpha},\ h_{1j}^{\alpha} = 0,\ \forall j \in \{2,...,n\}, \forall \alpha \in \{n+1,...,m\}, \\ h_{11}^{*\alpha} = h_{22}^{*\alpha} + ... + h_{nn}^{*\alpha},\ h_{1j}^{*\alpha} = 0,\ \forall j \in \{2,...,n\}, \forall \alpha \in \{n+1,...,m\}, \end{cases}$$

or, equivalently,

$$\begin{cases} 2h(X,X) = nH(p),\ h(X,Y) = 0,\ \forall Y \in T_pM^n \text{ orthogonal to } X, \\ 2h^*(X,X) = nH^*(p),\ h^*(X,Y) = 0,\ \forall Y \in T_pM^n \text{ orthogonal to X}. \end{cases}$$

Therefore, we proved the following Chen-Ricci inequality.

Theorem 2. *Let M^n be a statistical submanifold of a Hessian manifold $\tilde{M}^m(c)$ of constant Hessian curvature c. Then the Ricci curvature of a unit vector $X \in T_pM^n$ satisfies:*

$$Ric(X) \geq (n-1)\frac{c}{2} - \frac{n^2}{8}\|H\|^2 - \frac{n^2}{8}\|H^*\|^2 + Ric^0(X).$$

Moreover, the equality case holds if and only if

$$\begin{cases} 2h(X,X) = nH(p),\ h(X,Y) = 0,\ \forall Y \in T_pM^n \textit{ orthogonal to } X, \\ 2h^*(X,X) = nH^*(p),\ h^*(X,Y) = 0,\ \forall Y \in T_pM^n \textit{ orthogonal to } X. \end{cases}$$

Author Contributions: Both authors contributed equally to this research. The research was carried out by both authors, and the manuscript was subsequently prepared together. Both authors read and approved the final manuscript.

Conflicts of Interest: The authors declare no conflict of interest.

References

1. Chen, B.Y. Mean curvature and shape operator of isometric immersions in real-space-forms. *Glasg. Math. J.* **1996**, *38*, 87–97.
2. Chen, B.Y. Relations between Ricci curvature and shape operator for submanifolds with arbitrary codimensions. *Glasg. Math. J.* **1999**, *41*, 33–44.
3. Aydin, M.E.; Mihai, A.; Mihai, I. Some inequalities on submanifolds in statistical manifolds of constant curvature. *Filomat* **2015**, *29*, 465–477.
4. Opozda, B. A sectional curvature for statistical structures. *Linear Algebra Appl.* **2016**, 497, 134–161.
5. Aydin, M.E.; Mihai, A.; Mihai, I. Generalized Wintgen inequality for statistical submanifolds in statistical manifolds of constant curvature. *Bull. Math. Sci.* **2017**, *7*, 155–166.
6. Shima, H. *The Geometry of Hessian Structures*; World Scientific: Singapore, 2007.
7. Nomizu, K.; Sasaki, S. *Affine Differential Geometry*; Cambridge University Press: Cambridge, UK, 1994.
8. Amari, S. *Diffrential-Geometrical Methods in Statistics*; Springer: Berlin, Germany, 1985.
9. Simon, U. Affine Differential Geometry. In *Handbook of Differential Geometry*; Dillen, F., Verstraelen, L., Eds.; North-Holland: Amsterdam, The Netherlands, 2000; Volume I, pp. 905–961.

10. Vos, P.W. Fundamental equations for statistical submanifolds with applications to the Bartlett correction. *Ann. Inst. Stat. Math.* **1989**, *41*, 429–450.
11. Chen, B.Y. *Geometry of Submanifolds*; Marcel Dekker: New York, NY, USA, 1973.

mathematics

MDPI

Article

Completness of Statistical Structures

Barbara Opozda

Faculty of Mathematics and Computer Science, Jagiellonian University, ul. Łojasiewicza 6, 30-348 Cracow, Poland; Barbara.Opozda@im.uj.edu.pl

Received: 30 November 2018; Accepted: 11 January 2019; Published: 19 January 2019

Abstract: In this survey note, we discuss the notion of completeness for statistical structures. There are at least three connections whose completeness might be taken into account, namely, the Levi-Civita connection of the given metric, the statistical connection, and its conjugate. Especially little is known on the completeness of statistical connections.

Keywords: statistical structure; affine hypersurface; affine sphere; conjugate symmetric statistical structure; sectional ∇-curvature; complete connection

MSC: 53C05; 53A15

---

1. Introduction

In affine differential geometry, which is still the main source of statistical structures, the affine completeness of a nondegenerate affine hypersurface has always been meant as the completeness of the metric being the second fundamental form. In particular, Calabi's famous conjecture deals with affine completeness. Complete affine spheres are those whose Blaschke metric is complete, see, e.g., Reference [1]. This affine completeness has been opposed to Euclidean completeness, that is, completeness relative to the first fundamental form on a hypersurface. The completeness of the induced connections on affine hypersurfaces has never been studied. However, even if we restrict to the completeness of the second fundamental form and we switch from the geometry of affine hypersurfaces to the geometry of statistical structures, the situation becomes immediately much more complicated. It follows from the fact that, on a hypersurface, the induced structure has very strong properties that, in general, are not satisfied by an arbitrary statistical structure. In other words, not all statistical structures, even Ricci-symmetric, are realizable (even locally) on hypersurfaces. As examples of results on affine complete affine spheres, here we cite two classical theorems and three other theorems, being their analogs and generalizations in the category of statistical manifolds.

As for statistical connections, the first attempt to the study of their completeness was made by Noguchi [2]. He gave a procedure of constructing a complete statistical connection on a complete Riemannian manifold by using just one function. Statistical connections on compact manifolds are not necessarily complete. We provide a simple example on a torus. We also give a theorem generalizing the situation from this concrete simple example.

2. Preliminaries

We recall only those notions of statistical geometry that are needed in this note (for more information, see [3]). Let g be a positive definite Riemannian tensor field on a manifold M. Denote by $\hat\nabla$ the Levi-Civita connection for g. A statistical structure is a pair (g,∇), where ∇ is a torsion-free connection such that the following Codazzi condition is satisfied:

$$(\nabla_X g)(Y,Z) = (\nabla_Y g)(X,Z) \tag{1}$$

for all $X, Y, Z \in T_xM$, $x \in M$. A connection ∇ satisfying (1) is called a statistical connection for g. A statistical structure (g, ∇) is trivial if the statistical connection ∇ coincides with the Levi-Civita connection $\hat{\nabla}$.

For any connection ∇ one defines its conjugate connection $\overline{\nabla}$ relative to g by the following formula:

$$g(\nabla_X Y, Z) + g(Y, \overline{\nabla}_X Z) = Xg(Y, Z). \tag{2}$$

It is known that if (g, ∇) is a statistical structure, then so is $(g, \overline{\nabla})$. From now on, we assume that ∇ is a statistical connection for g.

If R is the curvature tensor for ∇, and $\overline{R}$ is the curvature tensor for $\overline{\nabla}$, then we have

$$g(R(X,Y)Z, W) = -g(\overline{R}(X,Y)W, Z). \tag{3}$$

Denote by Ric and $\overline{\mathrm{Ric}}$ the corresponding Ricci tensors. Note that, in general, these Ricci tensors are not symmetric. The curvature and the Ricci tensors of $\hat{\nabla}$ are denoted by $\hat{R}$ and $\widehat{\mathrm{Ric}}$. The function

$$\rho = \mathrm{tr}_g \mathrm{Ric}\,(\cdot,\cdot) \tag{4}$$

is called the scalar curvature of (g, ∇). Similarly, one can define the scalar curvature $\overline{\rho}$ for $(g, \overline{\nabla})$ but, by (3), $\rho = \overline{\rho}$. The function ρ is called the scalar statistical curvature. We also have the usual scalar curvature $\hat{\rho}$ for g.

We define the cubic form A by

$$A(X,Y,Z) = -\frac{1}{2}\nabla g(X,Y,Z), \tag{5}$$

where $\nabla g(X,Y,Z)$ stands for $(\nabla_X g)(Y,Z)$. It is clear that a statistical structure can be equivalently defined as a pair (g, A), where A is a symmetric cubic form.

The condition characterized by the following lemma plays a crucial role in our considerations.

Lemma 1. *Let (g, ∇) be a statistical structure. The following conditions are equivalent:*
(1) $R = \overline{R}$,
(2) $\hat{\nabla} A$ *is symmetric,*
(3) $g(R(X,Y)Z, W)$ *is skew-symmetric relative to* Z, W.

The family of statistical structures satisfying one of the above conditions is as important in the geometry of statistical structures as the family of affine spheres in affine differential geometry. A statistical structure satisfying Condition (2) in the above lemma was called conjugate symmetric in [4]. We adopt this name here. Note that condition $R = \overline{R}$ easily implies the symmetry of Ric.

A statistical structure is called trace-free if $\mathrm{tr}_g A(X,\cdot,\cdot) = 0$ for every $X \in TM$. This condition is equivalent to the condition that $\nabla \nu_g = 0$, where ν_g is the volume form determined by g.

In [3], we introduced the notion of the sectional ∇-curvature. Namely, the tensor field

$$\mathcal{R} = \frac{1}{2}(R + \overline{R}) \tag{6}$$

satisfies the following condition:

$$g(\mathcal{R}(X,Y)Z, W) = -g(\mathcal{R}(X,Y)W, Z).$$

If we denote by the same letter $\mathcal{R}$ the $(0,4)$-tensor field given by $\mathcal{R}(X,Y,W,Z) = g(\mathcal{R}(W,Z)Y,X)$, then this $\mathcal{R}$ has the same symmetries as the Riemannian $(0,4)$ curvature tensor. Therefore, we can define the sectional ∇-curvature by

$$k(\pi) = g(\mathcal{R}(e_1,e_2)e_2,e_1) \tag{7}$$

for a vector plane $\pi \in T_xM$, $x \in M$, where e_1, e_2 is any orthonormal basis of π. It is a well-defined notion, but it is not quite analogous to the Riemannian sectional curvature. For instance, in general, Schur's lemma does not hold for the sectional ∇-curvature. However, if a statistical structure is conjugate-symmetric (in this case $\mathcal{R} = R$) some type of the second Bianchi identity holds and, consequently, the Schur lemma holds [3].

The theory of affine hypersurfaces in $\mathbf{R}^{n+1}$ is a natural source of statistical structures. For the theory, we refer to [1] or [5]. We recall here only some basic facts.

Let $\mathbf{f}: M \to \mathbf{R}^{n+1}$ be a locally strongly convex hypersurface. For simplicity, assume that M is connected and orientable. Let ξ be a transversal vector field on M. The induced volume form ν_ξ on M is defined as follows:

$$\nu_\xi(X_1,...,X_n) = \det(\mathbf{f}_*X_1,...,\mathbf{f}_*X_n,\xi).$$

We also have the induced connection ∇ and second fundamental form g given by the Gauss formula:

$$D_X\mathbf{f}_*Y = \mathbf{f}_*\nabla_XY + g(X,Y)\xi,$$

where D is the standard flat connection on $\mathbf{R}^{n+1}$. Since the hypersurface is locally strongly convex, g is definite. By multiplying ξ by -1, if necessary, we can assume that g is positive definite. A transversal vector field is called equiaffine if $\nabla\nu_\xi = 0$. This condition is equivalent to the fact that ∇g is symmetric, i.e., (g,∇) is a statistical structure. It means, in particular, that for a statistical structure obtained on a hypersurface by a choice of an equiaffine transversal vector field, the Ricci tensor of ∇ is automatically symmetric. A hypersurface equipped with an equiaffine transversal vector field is called an equiaffine hypersurface.

Recall now the notion of the shape operator and the Gauss equations. Having a chosen equiaffine transversal vector field and differentiating it, we get the Weingarten formula:

$$D_X\xi = -\mathbf{f}_*\mathcal{S}X.$$

The tensor field $\mathcal{S}$ is called the shape operator for ξ. If R is the curvature tensor for the induced connection ∇, then

$$R(X,Y)Z = g(Y,Z)\mathcal{S}X - g(X,Z)\mathcal{S}Y. \tag{8}$$

This is the Gauss equation for R. The Gauss equation for $\overline{R}$ is the following:

$$\overline{R}(X,Y)Z = g(Y,\mathcal{S}Z)X - g(X,\mathcal{S}Z)Y. \tag{9}$$

It follows that the conjugate connection is projectively flat if $n > 2$. The conjugate connection is also projectively flat for two-dimensional surfaces equipped with an equiaffine transversal vector field, that is, that the cubic form $\overline{\nabla \mathrm{Ric}}$ is symmetric.

We have the volume form ν_g determined by g on M. In general, this volume form is not covariant constant relative to ∇. The central point of the classical affine differential geometry is the theorem saying that there is a unique equiaffine transversal vector field ξ, such that $\nu_\xi = \nu_g$. This unique transversal vector field is called the affine normal vector field or the Blaschke affine normal. The second fundamental form for the affine normal is called the Blaschke metric. A hypersurface endowed with the affine Blaschke normal is called a Blaschke hypersurface. Note that conditions $\nabla\nu_\xi = 0$ and $\nu_\xi = \nu_g$ imply that the statistical structure on a Blaschke hypersurface is trace-free.

If the affine lines determined by the affine normal vector field meet at one point or are parallel, then the hypersurface is called an affine sphere. In the first case, the sphere is called proper, in the second one improper. The class of affine spheres is very large. There exist a lot of conditions characterizing affine spheres. For instance, a hypersurface is an affine sphere if and only if $R = \overline{R}$. Therefore, conjugate symmetric statistical manifolds can be regarded as generalizations of affine spheres. For connected affine spheres, the shape operator S is a constant multiple of the identity, i.e., $S = k\,\mathrm{id}$. In particular, for affine spheres we have:

$$R(X,Y),Z = k\{g(Y,Z)X - g(X,Z)Y\}. \tag{10}$$

It follows that the statistical sectional curvature on a connected affine sphere is constant. If, as we have already done, we choose a positive definite Blaschke metric on a locally strongly convex affine sphere, then we call the sphere elliptic if $k > 0$, parabolic if $k = 0$, and hyperbolic if $k < 0$.

As we have already mentioned, if ∇ is a connection on a hypersurface induced by an equiaffine transversal vector field, then the conjugate connection $\overline{\nabla}$ is projectively flat. Therefore, the projective flatness of the conjugate connection is a necessary condition for (g, ∇) to be realizable as the induced structure on a hypersurface equipped with an equiaffine transversal vector field. In fact, roughly speaking, it is also a sufficient condition for local realizability. Note that, if (g, ∇) is a conjugate symmetric statistical structure, then ∇ and $\overline{\nabla}$ are simultaneously projectively flat. It follows that, if (g, ∇) is conjugate symmetric, then it is locally realizable on an equiaffine hypersurface if and only if ∇ or $\overline{\nabla}$ is projectively flat, and the realization is automatically on an affine sphere.

In [3,6], a few examples of conjugate symmetric statistical structures that are not realizable (even locally) on affine spheres were produced.

3. Statistical Structures with Complete Metrics

The following theorems are attributed to Blaschke, Deicke and Calabi (see e.g., [1]).

Theorem 1. *Let $f : M \to \mathbf{R}^{n+1}$ be an elliptic affine sphere whose Blaschke metric is complete. Then, M is compact and the induced structure on M is trivial. Consequently, the affine sphere is an ellipsoid.*

Theorem 2. *Let $f : M \to \mathbf{R}^{n+1}$ be a hyperbolic or parabolic affine sphere whose Blaschke metric is complete. Then, the Ricci tensor of the metric is negative semidefinite.*

The theorems can be generalized to the case of statistical manifolds in the following manner:

Theorem 3. *Let (g, ∇) be a trace-free conjugate symmetric statistical structure on a manifold M. Assume that g is complete on M. If the sectional ∇-curvature is bounded from below and above on M, then the Ricci tensor of g is bounded from below and above on M. If the sectional ∇-curvature is non-negative everywhere, then the statistical structure is trivial, that is, $\nabla = \hat{\nabla}$. If the statistical sectional curvature is bounded from 0 by a positive constant then, additionally, M is compact and its first fundamental group is finite.*

Let us explain why Theorem 3 is a generalization of Theorems 1 and 2. The induced structure on an affine sphere is a conjugate symmetric trace-free statistical structure. Moreover, the statistical connection on an affine sphere is projectively flat and its ∇-sectional curvature is constant. In Theorem 3, we do not need the projective flatness of the statistical connection, which means that the manifold with a statistical structure can be nonrealizable on any Blaschke hypersurface, even locally. Moreover, the assumption about the constant curvature is replaced by the assumption that the curvature satisfies some inequalities.

More precise and more general formulations of this theorem give the two following results:

Theorem 4. *Let (g, ∇) be a trace-free conjugate symmetric statistical structure on an n-dimensional manifold M. Assume that (M, g) is complete and the sectional ∇-curvature $k(\pi)$ satisfies the inequality*

$$H_3 + \frac{n-2}{2}\varepsilon \leq k(\pi) \leq H_3 + \frac{n}{2}\varepsilon \tag{11}$$

for every tangent plane π, where H_3 is a non-positive number and ε is a non-negative function on M. Then, the Ricci tensor $\widehat{\mathrm{Ric}}$ of g satisfies the following inequalities:

$$(n-1)H_3 + \frac{(n-1)(n-2)}{2}\varepsilon \leq \widehat{\mathrm{Ric}} \leq -(n-1)^2 H_3 + \frac{(n-1)n}{2}\varepsilon. \tag{12}$$

The scalar curvature $\hat{\rho}$ of g satisfies the following inequalities:

$$n(n-1)H_3 + \frac{(n-1)(n-2)n}{2}\varepsilon \leq \hat{\rho} \leq \frac{n^2(n-1)}{2}\varepsilon. \tag{13}$$

Theorem 5. *Let (M, g) be a complete Riemannian manifold with conjugate symmetric trace-free statistical structure (g, ∇). If the sectional ∇-curvature is non-negative on M, then the statistical structure is trivial, i.e., $\nabla = \hat{\nabla}$. Moreover, if the sectional ∇-curvature is bounded from 0 by a positive constant, then M is compact and its first fundamental group is finite.*

Proofs of Theorems 3–5 can be found in [6].

4. Completeness of Statistical Connections

Very tittle is known about the completeness of statistical connections. The difference between the completeness of metrics and that of affine connections is huge. In particular, a statistical connection on a compact manifold does not have to be complete. Indeed, we can offer the following simple example:

Example 1. *Take $\mathbf{R}^2$ with its standard flat Riemannian structure. Let U, V be the canonical frame field on $\mathbf{R}^2$. Define a statistical connection ∇ as follows:*

$$\nabla_U U = U, \ \ \nabla_U V = -V, \ \ \nabla_V V = -U. \tag{14}$$

This statistical structure can be projected on the standard torus T^2. A curve $\gamma(t) = (x(t), y(t))$ is a ∇-geodesic if and only if

$$\ddot{x} + (\dot{x})^2 - (\dot{y})^2 = 0, \ \ \ddot{y} - 2\dot{x}\dot{y} = 0. \tag{15}$$

Let y_0 be a fixed real number. Consider the curve

$$\gamma(t) = (\ln(1-t), y_0) \tag{16}$$

for $t \in [0, 1)$. It is a ∇-geodesic. We have $\|\dot{\gamma}(t)\| = \frac{1}{1-t} \to +\infty$ if $t \to 1$. Hence, this geodesic cannot be extended beyond 1. The connection ∇ is not complete on T^2.

In the above example, the cubic form A of the statistical structure is $\hat{\nabla}$-parallel. This is the reason why the statistical connection is not complete. More precisely, we have:

Theorem 6. ([7]) *Let (g, ∇) be a non-trivial statistical structure such that*

$$\hat{\nabla} A(U, U, U, U) \leq 0$$

for every $U \in \mathcal{U}M$, where $\mathcal{U}M$ is the unit sphere bundle over M. The statistical connection ∇ is not complete.

As a corollary, we get:

Corollary 1. *Let (g, ∇) be a statistical structure for which ∇ is complete and $(\hat{\nabla} A)(U, U, U, U) \leq 0$ for each $U \in \mathcal{U}M$. Then, the statistical structure must be trivial.*

Let us now cite a positive result first proved by Noguchi [2].

Theorem 7. *Let (M, g) be a complete Riemannian manifold, and A be a cubic form given by:*

$$A = sym(d\sigma \otimes g) \tag{17}$$

for some function σ on M. Assume that the function σ is bounded from below on M. Then, the statistical connection of statistical structure (g, A) is complete.

In particular, any function on a compact Riemannian manifold M gives rise to a statistical structure on M whose statistical connection is complete. In fact, we have a more general fact:

Corollary 2. *Let (M, g) be a compact Riemannian manifold. Each function σ on M gives rise to a statistical structure whose statistical connection and its conjugate are complete.*

Funding: This research received no external funding.

Conflicts of Interest: The author declares no conflict of interest.

References

1. Li, A.-M.; Simon, U.; Zhao, G. *Global Affine Differential Geometry of Hypersurfaces*; W. de Greuter: Berlin, Germany; New York, NY, USA, 1993.
2. Noguchi, M. Geometry of statistical manifolds. *Differ. Geom. Appl.* **1992**, *2*, 197–222. [CrossRef]
3. Opozda, B. Bochner's technique for statistical structures. *Ann. Glob. Anal. Geom.* **2015**, *48*, 357–395. [CrossRef]
4. Lauritzen S.L. *Statistical Manifolds*; IMS Lecture Notes-Monograph Series 10; Institute of Mathematical Statistics: Beachwood, OH, USA, 1987; pp. 163–216.
5. Nomizu, K.; Sasaki, T. *Affine Differential Geometry, Geometry of Affine Immersions*; Cambridge University Press: Cambridge, UK, 1994.
6. Opozda, B. Curvature bounded conjugate symmetric statistical structures with complete metric. *arXiv*, **2018**, arXiv:1805.07807.
7. Opozda, B. Complete statistical connections. 2019, preprint.

mathematics

MDPI

Article

L^2-Harmonic Forms on Incomplete Riemannian Manifolds with Positive Ricci Curvature †

Junya Takahashi

Research Center for Pure and Applied Mathematics, Graduate School of Information Sciences, Tôhoku University, 6-3-09, Aoba, Sendai 980-8579, Japan; t-junya@tohoku.ac.jp
† Dedicated to the Memory of Professor Ahmad El Soufi.

Received: 28 February 2018; Accepted: 29 April 2018; Published: 9 May 2018

Abstract: We construct an incomplete Riemannian manifold with positive Ricci curvature that has non-trivial L^2-harmonic forms and on which the L^2-Stokes theorem does not hold. Therefore, a Bochner-type vanishing theorem does not hold for incomplete Riemannian manifolds.

Keywords: L^2-harmonic forms; Hodge–Laplacian; manifold with singularity; L^2-Stokes theorem; capacity

MSC: Primary 58A14; Secondary 58A12, 14F40, 53C21

---

1. Introduction

The Stokes theorem or the Green formula plays a very important role in geometry and analysis on manifolds. For example, we recall the proof of the Bochner vanishing theorem (e.g., [1] p. 185, Theorem 4.5.2).

Theorem 1 (Bochner vanishing theorem). *Let (M, g) be a connected oriented closed Riemannian manifold. If the Ricci curvature Ric > 0 on M, then the first cohomology group $H^1(M; \mathbb{R}) = 0$.*

From the proof of the Bochner vanishing theorem, it follows that, if the Stokes theorem does not hold on an incomplete Riemannian manifold of positive Ricci curvature, then the Bochner vanishing theorem for it might not hold. It is a natural question to ask whether or not the Stokes theorem on general incomplete Riemannian manifolds holds. Indeed, Cheeger in [2] studied the Stokes theorem and the Hodge theory on Riemannian manifolds with conical singularities, more generally, Riemannian pseudomanifolds. The analysis on pseudomanifolds is, by definition, the L^2-analysis on the regular set that excludes the singular points. Then, there are many valuable results on Riemannian pseudomainfolds (e.g., [3,4]). Indeed, Cheeger, Goresky and MacPherson in [4] stated that the L^2-cohomology groups of the regular sets of Riemannian pseudomanifolds are isomorphic to the intersection cohomology groups with the lower middle perversities. These studies have still been developing by many mathematicians (see [5–8]). Recently, Albin, Leichtnam, Mazzeo and Piazza in [9] studied the Hodge theory on more general singular spaces, which were called Cheeger spaces.

On the other hand, Cheeger ([2] p. 140, Theorem 7.1 and [10] p. 34, Theorem 3) proved that generalized Bochner-type vanishing theorems hold on some Riemannian pseudomanifolds with a kind of "positive curvature". This kind of "positive curvature" seems to behave like a positive curvature operator.

However, it seems that there are no concrete examples where a Bochner-type vanishing theorem does not hold. Thus, we construct a simple concrete example where a Bochner-type vanishing theorem does not hold. Note that a Bochner-type vanishing theorem holds for complete Riemannian manifolds [11].

In the present paper, we give an incomplete Riemannian manifold with positive Ricci curvature for which a Bochner-type vanishing theorem does not hold. The construction of our manifold is the following way. Let (N^n, h) be a connected oriented closed Riemannian manifold of dimension n. We consider the suspension $\Sigma(N)$ of N, and equip the smooth set of $\Sigma(N)$ with a Riemannian metric g. We denote by $\overline{M}$ the suspension of N:

$$\overline{M} := \Sigma(N) = [0, \pi] \times N \big/ \sim,$$

where the equivalent relation is

$$(r_1, y_1) \sim (r_2, y_2) \overset{\text{equiv.}}{\Longleftrightarrow} r_1 = r_2 = 0 \text{ or } \pi$$

for $(r_1, y_1), (r_2, y_2) \in [0, \pi] \times N$. Let $M = \overline{M}_{\text{reg}}$ be the regular set of $\overline{M}$, which consists of all smooth points of $\overline{M}$, i.e., $\overline{M}_{\text{reg}} = (0, \pi) \times N$. The singular set is $\overline{M}_{\text{sing}} := \overline{M} \setminus \overline{M}_{\text{reg}}$, i.e., two vertices corresponding to $r = 0, \pi$. We define an incomplete Riemannian metric g on this smooth part $M = (0, \pi) \times N$ as

$$g := dr^2 \oplus \sin^{2a}(r)h$$

for some constant $0 < a < 1$. In fact, we take $a = \frac{1}{n}$. This metric is a warped product metric with the warping function $\sin^a(r)$. Then, our main theorem is stated as follows:

Theorem 2. *There exists an incomplete Riemannian manifold (M^m, g) of dimension $m \geq 2$ satisfying the following four properties:*

(1) *the Ricci curvature of (M, g) is $\mathrm{Ric} \geq K > 0$ for some constant $K > 0$;*
(2) *there exist non-trivial L^2-harmonic p-forms on (M, g) for all $1 \leq p \leq m - 2$;*
(3) *the L^2-Stokes theorem for all $1 \leq p \leq m - 2$ does not hold on (M, g);*
(4) *the capacity of the singular set satisfies $Cap(\overline{M}_{sing}) = 0$.*

Remark 1. (*i*) *In the case of p =1, Theorem 2 implies that a Bochner-type vanishing theorem does not hold for an incomplete Riemannian manifold with $Ric \geq K > 0$.*

(*ii*) *The curvature operator on (M, g) is not positive. However, we do not know whether or not the Weitzenböck curvature tensor F_p is positive, where F_p is the curvature term in the Weitzenböck formula for p-form φ:*

$$-\frac{1}{2}\Delta(|\varphi|_g^2) = -\langle \Delta\varphi, \varphi \rangle_g + |\nabla\varphi|_g^2 + \langle F_p\varphi, \varphi \rangle_g. \tag{1}$$

Therefore, we do not apply the Bochner-type vanishing theorem for all p-forms by Gallot and Meyer [12], p. 262, Proposition 0.9. Note that the Weitzenböck curvature tensor is estimated below by a lower bound of the curvature operator (e.g., [13], p. 346, Corollary 9.3.4).

(*iii*) *For harmonic 1-form $\varphi = d\theta$ on $\mathbb{T}^n$, by the Equation (1) and F_1 = Ric, there exists non-constant subharmonic function $|d\theta|_g^2 = sin^{-2/n}(r)$ on $M = (0, \pi) \times \mathbb{T}^n$, that is, $\Delta(|\varphi|_g^2) \leq 0$ on M.*

The present paper is organized as follows: In Section 2, we recall two important closed extensions of the exterior derivative d, which are $d_{\max}$ and $d_{\min}$, and the L^2-Stokes theorem on Riemannian manifolds with conical singularity by Cheeger [2]. In Section 3, we calculate L^2-harmonic forms on a warped product Riemannian manifold and the capacity of the vertex. In Section 4, the final section, we prove Theorem 2.

2. L^2-Stokes Theorem

Let (M^m, g) be a connected oriented (possibly incomplete) Riemannian manifold of dimension m. We denote by $\Omega_0^p(M)$ the set of all smooth p-forms on M with compact support, and by d_p the exterior derivative acting on smooth p-forms. We consider the de Rham complex $d_p : \Omega_0^p(M) \longrightarrow \Omega_0^{p+1}(M)$ for $p = 0, 1, 2, \ldots, m-1$ with $d_{p+1} \circ d_p \equiv 0$. By using the Riemannian metric g, we define the L^2-inner product on $\Omega_0^p(M)$ as

$$(\varphi, \psi)_{L^2(\Lambda^p M, g)} := \int_M \langle \varphi, \psi \rangle_g \, d\mu_g$$

for any $\varphi, \psi \in \Omega_0^p(M)$, where $d\mu_g$ is the Riemannian measure and $\langle \ , \ \rangle_g$ is the fiber metric on the exterior bundle $\Lambda^p T^* M$ induced from the Riemannian metric g. The space of L^2 p-forms $L^2(\Lambda^p M, g)$ is the completion of $\Omega_0^p(M)$ with respect to this L^2-norm.

Next, we consider the completion of the exterior derivative d_p, which induces a Hilbert complex introduced by Brüning and Lesch [14], p. 90. (See also Bei [5], pp.6–8). There are two important closed extensions of d_p, one of which is the maximal extension $d_{p,\max}$ and the other is the minimal extension $d_{p,\min}$.

Definition 1 (maximal extension $d_{p,\max}$). *The maximal extension $d_{p,\max}$ is the operator acting on the domain:*

$$\mathrm{Dom}(d_{p,\max}) := \Big\{ \varphi \in L^2(\Lambda^p M, g) \,\Big|\, \textit{There exists } \psi \in L^2(\Lambda^{p+1} M, g) \textit{ such that } (\varphi, \delta_{p+1}\eta)_{L^2(\Lambda^p M,g)} = (\psi, \eta)_{L^2(\Lambda^{p+1} M,g)} \textit{ for any } \eta \in \Omega_0^{p+1}(M) \Big\},$$

and, in this case, we write

$$d_{p,\max}\varphi = \psi.$$

In other words, $\mathrm{Dom}(d_{p,\max})$ *is the largest set of differential p-forms $\varphi \in L^2(\Lambda^p M, g)$ such that the distributional derivative $d_p\varphi$ is also in $L^2(\Lambda^{p+1} M, g)$.*

Definition 2 (minimal extension $d_{p,\min}$). *The minimal extension $d_{p,\min}$ is given by the closure with respect to the graph norm of d_p in $L^2(\Lambda^p M, g)$, that is,*

$$Dom(d_{p,min}) := \Big\{ \varphi \in L^2(\Lambda^p M, g) \,\Big|\, \textit{There exists } \{\varphi_i\}_i \in \Omega_0^p(M) \textit{ such that } \varphi_i \to \varphi,\ d_p\varphi_i \to \psi \in L^2(\Lambda^{p+1} M, g)\ (L^2\textit{-strongly}) \Big\},$$

and, in this case, we write

$$d_{p,\min}\varphi = \psi.$$

In other words, $d_{p,\min}$ is the smallest closed extension of d_p, that is, $d_{p,\min} = \overline{d_p}$.

It is obvious that

$$\Omega_0^p(M) \subset \mathrm{Dom}(d_{p,\min}) \subset \mathrm{Dom}(d_{p,\max}).$$

In the same manner, from the co-differential operator $\delta_p := (-1)^{mp+m+1} * d_{m-p} * : \Omega_0^p(M) \longrightarrow \Omega_0^{p-1}(M)$, where $*$ is the Hodge $*$-operator on (M, g), we can define the maximal extension $\delta_{p,\max}$ and the minimal extension $\delta_{p,\min}$. These operators are mutually adjoint, that is,

$$(\delta_{p+1,\min})^* = d_{p,\max}, \quad (\delta_{p+1,\max})^* = d_{p,\min}. \tag{2}$$

Note that min and max are exchanged.

Now, we recall the definition of the L^2-Stokes theorem for p-forms (see Cheeger [2] p. 95 (1,7), [15] p. 72, Definition 2.2, [16] p. 40, Definition 4.1).

Definition 3 (L^2-Stokes theorem)**.** *Let (M^m, g) be a connected oriented Riemannian manifold. The L^2-Stokes theorem for p-forms holds on (M, g), if*

$$(d_{p,max}\varphi, \psi)_{L^2(\Lambda^{p+1} M,g)} = (\varphi, \delta_{p+1,max}\psi)_{L^2(\Lambda^p M,g)} \tag{3}$$

for any $\varphi \in Dom(d_{p,max})$ and $\psi \in Dom(\delta_{p+1,max})$.

For complete Riemannian manifolds, the L^2-Stokes theorem for all p-forms always holds (Gaffney [17,18]).

Since the Equation (3) implies $d_{p,\max} = (\delta_{p+1,\max})^*$, the L^2-Stokes theorem for p-forms holds if and only if $d_{p,\min} = d_{p,\max}$, i.e., a closed extension of d_p is unique.

Now, for any $\varphi \in \mathrm{Dom}(d_{p,\max})$ and $\psi \in \mathrm{Dom}(\delta_{p+1,\max})$, we see that

$$\begin{aligned}(d_{\max}\varphi,\ \psi)_{L^2(\Lambda^{p+1} M,g)} - (\varphi,\ \delta_{\max}\psi)_{L^2(\Lambda^p M,g)} &= \int_M \langle d_{\max}\varphi,\ \psi\rangle d\mu_g - \int_M \langle \varphi,\ \delta_{\max}\psi\rangle d\mu_g \\ &= \int_M d_{L^1,\max}(\varphi \wedge *_g\psi),\end{aligned}$$

where the last $d_{L^1,\max}$ is the maximal extension of d_{m-1} between $L^1(\Lambda^* M, g)$, that is, the domain is $\{\, \omega \in L^1(\Lambda^{m-1} M, g) \mid d\omega \in L^1(\Lambda^m M, g)$ (in the distribution sense) $\}$. Therefore, we have

Lemma 1. *The L^2-Stokes theorem for p-forms holds on (M, g) if and only if*

$$\int_M d_{L^1,max}(\varphi \wedge *_g\psi) = 0$$

for any $\varphi \in Dom(d_{p,max})$ and $\psi \in Dom(\delta_{p+1,max})$.

Remark 2. *Gaffney ([18] p. 141, Theorem) proved the L^1-Stokes theorem, or the special Stokes theorem, for oriented complete Riemannian manifolds: If any smooth $(m-1)$-form ω on an oriented complete Riemannian manifold of dimension m such that $\omega, d\omega$ are in $L^1(\Lambda^* M, g)$, then*

$$\int_M d\omega = 0.$$

This L^1-Stokes theorem implies the L^2-Stokes theorem for all p-forms, but the inverse does not hold (see Grigor'yan and Masamune [19] p. 614, Proposition 2.4).

We recall connected oriented compact Riemannian manifolds with conical or horn singularity (Cheeger [2,3]). Let (N^n, h) be a connected oriented closed Riemannian manifold of dimension n, and let M_1^m be a connected oriented compact manifold of dimension $m = n + 1$ with the boundary

$\partial M_1 = N$. Let $f : I = [0, l] \longrightarrow \mathbb{R}_+$ be a smooth function with $f(0) = 0$ and $f(r) > 0$ for $r > 0$. The metric f-horn $C_f(N)$ over (N, h) is defined as the metric space

$$C_f(N) = I \times N \big/ \sim,$$

where the equivalent relation is

$$(r_1, y_1) \sim (r_2, y_2) \overset{\text{equiv.}}{\Longleftrightarrow} r_1 = r_2 = 0$$

for $(r_1, y_1), (r_2, y_2) \in I \times N$. The Riemannian metric g_f on the regular set $C_f(N)_{\text{reg}} = (0, l] \times N$ is defined as

$$g_f := dr^2 \oplus f^2(r)h \quad \text{on} \quad (0, l] \times N.$$

Then, we glue M_1 to $C_f(N)$ along their boundary N, and the resulting manifold denotes $M := M_1 \cup_N C_f(N)$. We introduce a smooth Riemannian metric g on the regular part $M_{\text{reg}} = M_1 \cup_N C_f(N)_{\text{reg}}$ such that g smoothly extends to M_1 from the f-horn metric g_f on $C_f(N)_{\text{reg}} = (0, l] \times N$. Thus, we obtain a connected oriented compact Riemannian manifold with f-horn singularity

$$(M^m, g) = (M_1, g) \cup_N (C_f(N), g_f).$$

Then, Cheeger proved the L^2-Stokes theorem on a compact Riemannian manifold with f-horn singularity.

Theorem 3. *We use the same notation as above. Let $(M^m, g) = (M_1, g) \cup_N (C_f(N), g_f)$ be a connected oriented compact Riemannian manifold with f-horn singularity. Suppose that the function $f(r) = r^a$ with positive constant $a \geq 1$. Then, for a compact Riemannian manifold with r^a-horn singularity (M^m, g), the following hold [Cheeger [2]] :*

(1) *If $n = 2k + 1$, the L^2-Stokes theorem holds for all p-forms on (M, g);*
(2) *If $n = 2k$, the L^2-Stokes theorem holds for all p-forms except $p = k$ on (M, g);*
(3) *If $n = 2k$, and if $H^k(N; \mathbb{R}) = 0$, the L^2-Stokes theorem holds for k-forms on (M, g);*
(4) *If $n = 2k$, and if $H^k(N; \mathbb{R}) \neq 0$, the L^2-Stokes theorem does not hold for k-forms on (M, g).*

Thus, Cheeger gave a necessary and sufficient condition that the L^2-Stokes theorem holds on a compact Riemannian manifold with r^a-horn singularity for $a \geq 1$.

Moreover, when $n = 2k$, Brüning and Lesch [20] p. 453, Theorem 3.8, gave a choice of ideal boundary conditions. More precisely,

Theorem 4. *In the case of $a = 1$ as in Theorem 3 [Brüning and Lesch [20]], we have*

$$\mathrm{Dom}(d_{p,\max}) \big/ \mathrm{Dom}(d_{p,\min}) \cong \begin{cases} H^k(N; \mathbb{R}), & \text{if } n = 2k \text{ and } p = k, \\ 0, & \text{otherwise.} \end{cases}$$

Remark 3. (*i*) *Since $\dim H^k(N; \mathbb{R})$ is finite, closed extensions of $d_{p,\min}$ are at most finite.*

(*ii*) *In the case of more complicated singularities, Hunsicker and Mazzeo [21] proved the L^2-Stokes theorem on Riemannian manifolds with edges (see [21] p. 3250, Corollary 3.11, or [16] p. 64, Theorem 5.11).*

3. Warped Product Manifolds

We consider L^2-harmonic forms, the Ricci curvature, and the capacity of the Cauchy boundary for a general warped product Riemannian manifold.

Let (N^n, h) be a connected oriented closed Riemannian manifold of dimension n. Let $f : (0, l) \longrightarrow \mathbb{R}_+$ be a smooth positive function with $f(+0) = 0$. Suppose that $f(r)$ is the same order of r^a for some constant $0 < a < 1$, that is, there exists a positive constant $C > 0$ such that

$$C^{-1}r^a \leq f(r) \leq Cr^a \quad (0 < r < l).$$

Then, we consider the warped product Riemannian manifold

$$M_f = (M^m, g) := ((0, l) \times N,\, dr^2 \oplus f(r)^2 h)$$

of dimension $m := \dim M_f = n + 1$. This Riemannian manifold (M, g) is incomplete at $r = +0$. We denote by x_0 the vertex of the f-horn $C_f(N)$ corresponding to $r = 0$.

Now, we can naturally extend p-forms on N to the p-forms on $M = (0, l) \times N$: $\Omega^p(N) \subset \Omega^p(M)$.

Lemma 2. *For any harmonic p-form φ on (N, h), the natural extension φ on M is also a harmonic p-form on (M, g).*

Proof. First, we have $d_M\varphi = d_N\varphi = 0$ on M. Next, it is easy to see that

$$*_g(\varphi) = (-1)^p f(r)^{n-2p} dr \wedge *_h(\varphi).$$

Hence, since $d_N(*_h(\varphi)) = 0$ by the harmonicity of φ on (N, h), we have

$$\begin{aligned} d_M(*_g\varphi) &= (-1)^p\, d_M\big(f(r)^{n-2p} dr \wedge *_h(\varphi) \big) \\ &= (-1)^{p+1} f(r)^{n-2p} dr \wedge d_N\big(*_h \varphi \big) = 0. \end{aligned}$$

Therefore, we find that φ is harmonic on (M, g) □

Lemma 3. *If $p < \frac{1}{2}(n + \frac{1}{a})$, then any smooth p-form φ on N naturally extends to $L^2(\Lambda^p M, g)$.*

Proof. For any $\varphi \in \Omega^p(N)$, we have

$$\begin{aligned} \|\varphi\|^2_{L^2(\Lambda^p M, g)} &= \int_0^l \int_N |\varphi|_g^2\, d\mu_g = \int_0^l \int_N |\varphi|^2_{f^2h} f(r)^n\, dr d\mu_h \\ &= \int_0^l f^{n-2p}(r) dr \int_N |\varphi|_h^2\, d\mu_h \leq C^{n-2p} \int_0^l r^{a(n-2p)}\, dr \|\varphi\|^2_{L^2(\Lambda^p N, h)}. \end{aligned}$$

Since $a(n - 2p) > -1$, the integral $\int_0^l r^{a(n-2p)}\, dr$ converges. Thus, we find $\varphi \in L^2(\Lambda^p M, g)$. □

Now, we take a cut-off function $\chi \in C^\infty(M)$ such that

$$\chi(r) := \begin{cases} 1, & \text{if } r \leq \frac{l}{4}, \\ 0, & \text{if } \frac{l}{2} \leq r. \end{cases}$$

If we set

$$\widetilde{\varphi} := \chi(r)\varphi \ \text{ on } \ M = (0, l) \times N, \tag{4}$$

then we see that $\widetilde{\varphi} \in \Omega^p(M)$ and the support $\mathrm{supp}(\widetilde{\varphi}) \subset (0, \frac{l}{2}] \times N$.

Lemma 4. *For any harmonic p-form $\varphi \in \Omega^p(N)$, the p-form $\widetilde{\varphi}$ on M satisfies*

(1) $\quad \widetilde{\varphi} \in Dom(d_{p,max})$, *if* $p < \dfrac{1}{2}\Big(n + \dfrac{1}{a}\Big)$;

(2) $\quad f(r)^{2p-n}dr \wedge \widetilde{\varphi} \in Dom(\delta_{g_{p+1,max}})$, *if* $p > \frac{1}{2}\left(n - \frac{1}{a}\right)$.

Proof. (1) First, since $p < \frac{1}{2}(n + \frac{1}{a})$, by Lemma 3, the p-form $\widetilde{\varphi} \in \mathrm{Dom}(d_{p,\max})$ is in $L^2(\Lambda^p M, g)$. Next, since $d_N\varphi = 0$ by the harmonicity of φ on (N, h), then we have

$$d\widetilde{\varphi} = d(\chi\varphi) = d\chi \wedge \varphi + \chi d_N\varphi = \chi'(r)dr \wedge \varphi \ \text{ on } \left[\tfrac{l}{4}, \tfrac{l}{2}\right] \times N.$$

Hence, since

$$\|d\widetilde{\varphi}\|^2_{L^2(\Lambda^{p+1} M,g)} = \|d\widetilde{\varphi}\|^2_{L^2(\Lambda^{p+1}[\frac{l}{4},\frac{l}{2}]\times N,g)} < \infty,$$

we see that $d\widetilde{\varphi} \in L^2(\Lambda^{p+1} M, g)$. Thus, we find $\widetilde{\varphi} \in \mathrm{Dom}(d_{p,\max})$.

(2) We prove $f(r)^{2p-n}dr \wedge \widetilde{\varphi} \in \mathrm{Dom}(\delta_{g_{p+1,\max}})$, if $p > \frac{1}{2}(n - \frac{1}{a})$. It is easy to see that

$$*_g\left(f(r)^{2p-n}dr \wedge \widetilde{\varphi}\right) = *_h(\widetilde{\varphi}). \tag{5}$$

Since $*_h(\varphi) \in \Omega^{n-p}(N)$ and $n - p < \frac{1}{2}(n + \frac{1}{a})$, by Lemma 3, we see $*_h(\varphi) \in L^2(\Lambda^{n-p} M, g)$. Thus, from the Equation (5), it follows that

$$\|f(r)^{2p-n}dr \wedge \widetilde{\varphi}\|^2_{L^2(\Lambda^{p+1} M,g)} = \| *_h(\widetilde{\varphi})\|^2_{L^2(\Lambda^{n-p} M,g)} \le \| *_h(\varphi)\|^2_{L^2(\Lambda^{n-p} M,g)} < \infty.$$

Hence, we see $f(r)^{2p-n}dr \wedge \widetilde{\varphi} \in L^2(\Lambda^{p+1} M, g)$.
Next, since $d_N(*_h\varphi) \equiv 0$ by the harmonicity of φ on (N, h), we have

$$d_M(*_h\widetilde{\varphi}) = d_M(\chi *_h(\varphi)) = \chi' dr \wedge (*_h\varphi). \tag{6}$$

Hence, from the proof of Lemma 4 (1), it follows that

$$\begin{aligned}
\|\delta_g(f(r)^{2p-n}dr \wedge \widetilde{\varphi})\|^2_{L^2(\Lambda^p M,g)} &= \|d *_g (f(r)^{2p-n}dr \wedge \widetilde{\varphi})\|^2_{L^2(\Lambda^{m-p} M,g)} \\
&= \|d *_h(\widetilde{\varphi})\|^2_{L^2(\Lambda^{m-p} M,g)} \quad \text{(by the Equation (5))} \\
&= \|\chi' dr \wedge (*_h\varphi)\|^2_{L^2(\Lambda^{m-p} M,g)} \quad \text{(by the Equation (6))} \\
&= \|\chi' dr \wedge (*_h\varphi)\|^2_{L^2(\Lambda^{m-p}[\frac{l}{4},\frac{l}{2}]\times N,g)} < \infty.
\end{aligned}$$

Therefore, we find $f(r)^{2p-n}dr \wedge \widetilde{\varphi} \in \mathrm{Dom}(\delta_{g_{p+1,\max}})$. □

If we make good choices of N and a, we have the following lemma.

Lemma 5. *If $H^p(N; \mathbb{R}) \neq 0$ for some p satisfying $\frac{1}{2}(n - \frac{1}{a}) < p < \frac{1}{2}(n + \frac{1}{a})$, then the L^2-Stokes theorem for p-forms does not hold on (M, g).*

Proof. Since $H^p(N, \mathbb{R}) \neq 0$, by the de Rham–Hodge–Kodaira theory, there exists a non-zero harmonic p-form $\varphi \neq 0$ on N. From Lemma 4, it follows that $\widetilde{\varphi} \in \mathrm{Dom}(d_{\max,p})$ and that $f(r)^{2p-n}dr \wedge \widetilde{\varphi} \in \mathrm{Dom}(\delta_{g_{\max,p+1}})$. Then, by the Equation (5), we have

$$\widetilde{\varphi} \wedge *_g(f(r)^{2p-n}dr \wedge \widetilde{\varphi}) = \widetilde{\varphi} \wedge *_h(\widetilde{\varphi}) = \chi^2(r)|\varphi|^2_h v_h,$$

where v_h is the volume form of (N, h). Since $\chi \equiv 1$ on $(0, \frac{l}{4}] \times N$, we have

$$\int_M d(\widetilde{\varphi} \wedge *_g (f(r)^{2p-n} dr \wedge \widetilde{\varphi})) = \int_M d(\chi^2(r)|\varphi|_h^2 v_h)$$
$$= \int_{(0,\frac{l}{4}] \times N} d(|\varphi|_h^2 v_h) + \int_{[\frac{l}{4},\frac{l}{2}] \times N} d(\chi^2(r)|\varphi|_h^2 v_h).$$

Since $d(|\varphi|_h^2 v_h)$ is an $(n+1)$-form on N^n, the first term is 0. Next, by the usual Stokes theorem, the second term is

$$\int_{[\frac{l}{4},\frac{l}{2}] \times N} d(\chi^2(r)|\varphi|_h^2 v_h) = \int_{\{\frac{l}{2}\} \times N} \chi^2(\tfrac{l}{2})|\varphi|_h^2 v_h - \int_{\{\frac{l}{4}\} \times N} \chi^2(\tfrac{l}{4})|\varphi|_h^2 v_h$$
$$= -\int_{\{\frac{l}{4}\} \times N} |\varphi|_h^2 v_h \;\; (\text{since } \chi(\tfrac{l}{4}) = 1,\ \chi(\tfrac{l}{2}) = 0)$$
$$= -\|\varphi\|_{L^2(\Lambda^p N, h)}^2 \neq 0.$$

Therefore, we have

$$\int_M d(\widetilde{\varphi} \wedge *_g (f^{2p-n}(r) dr \wedge \widetilde{\varphi})) \neq 0.$$

From Lemma 1, the L^2-Stokes theorem for p-forms does not hold on (M, g). □

Now, we recall the Ricci curvature of a warped product Riemannian manifold (M, g) (e.g., [22], p. 266, Proposition 9.106).

Lemma 6 (Ricci curvature). *Let $\{e_1, \ldots, e_n\}$ be a local orthonormal frame of (N^n, h). We set the local orthonormal local frame of (M, g) as $\{\widetilde{e}_0 := \frac{\partial}{\partial r}, \widetilde{e}_1 := f^{-1}e_1, \ldots, \widetilde{e}_n := f^{-1}e_n\}$. Then, the Ricci operator on (M^{n+1}, g) is given by*

(1) $$Ric_g(\widetilde{e}_0) = -n \frac{f''(r)}{f(r)} \widetilde{e}_0;$$

(2) $$Ric_g(\widetilde{e}_i) = Ric_h(\widetilde{e}_i) - \left\{ \frac{f''(r)}{f(r)} + (n-2) \left(\frac{f'(r)}{f(r)} \right)^2 \right\} \widetilde{e}_i, \quad (i = 1, \ldots, n).$$

We recall the definition of the capacity of a subset (see [23] **2.1** pp. 64–65 or [19] p. 612).

Definition 4 (capacity). *For any open subset $U \subset M$, the capacity, or 1-capacity, of U is defined as*

$$Cap(U) := inf \left\{ \|u\|_{H^1(M,g)}^2 \,\middle|\, u \in H^1(M, g) \text{ and } u \geq 1 \text{ a.e. } U \right\},$$

where $\|u\|_{H^1(M,g)}^2 = \|u\|_{L^2(M,g)}^2 + \|du\|_{L^2(\Lambda^* M,g)}^2$ is the Sobolev norm of u in the Sobolev space $H^1(M, g)$. If there exist no such functions, then we define $Cap(U) := \infty$. For any subset $A \subset M$, we define

$$Cap(A) := inf \left\{ Cap(U) \,\middle|\, \text{any open subset } U \text{ with } A \subset U \subset M \right\}.$$

Now, we compute the capacity of the Cauchy boundary $\partial_c M := \overline{M} \setminus M = \{x_0\}$, where $\overline{M}$ is the completion as the metric space M with respect to the Riemannian distance d_g.

Lemma 7. *If $a \geq \frac{1}{n}$, then we have $Cap(\partial_c M) = 1$.*

Proof. We take the cut-off function $\chi_\varepsilon : [0,l) \to [0,1]$ such that

$$\chi_\varepsilon(r) := \begin{cases} 1, & (0 \le r \le \varepsilon), \\ 1 + \dfrac{2}{\log \varepsilon} \log\left(\dfrac{r}{\varepsilon}\right), & (\varepsilon \le r \le \sqrt{\varepsilon}), \\ 0, & (\sqrt{\varepsilon} \le r). \end{cases} \tag{7}$$

Set $\chi_\varepsilon(x) := \chi_\varepsilon(d_g(x_0, x))$ for $x \in M$. Then, $\chi_\varepsilon \in H^1(M,g)$ and $|\chi_\varepsilon| \le 1$ on the geodesic ball of radius $\sqrt{\varepsilon} > 0$ centered at x_0.

We prove that $\|\chi_\varepsilon\|^2_{L^2(M,g)} \to 0$ as $\varepsilon \to 0$. First, it is easy to see that

$$\begin{aligned} \|\chi_\varepsilon\|^2_{L^2(M,g)} &= \int_M |\chi_\varepsilon(r)|^2 d\mu_g = \int_0^{\sqrt{\varepsilon}} |\chi_\varepsilon(r)|^2 f(r)^n dr \int_N d\mu_h \\ &\le \int_0^{\sqrt{\varepsilon}} f(r)^n dr \operatorname{vol}(N,h) \le C^n \operatorname{vol}(N,h) \int_0^{\sqrt{\varepsilon}} r^{na} dr \\ &\le C^n \operatorname{vol}(N,h) \int_0^{\sqrt{\varepsilon}} 1\, dr \quad (\text{by } na \ge 1) \\ &= C^n \operatorname{vol}(N,h) \sqrt{\varepsilon} \longrightarrow 0 \quad (\text{as } \varepsilon \to 0). \end{aligned} \tag{8}$$

Next, we prove that $\|d\chi_\varepsilon\|^2_{L^2(\Lambda^1 M,g)} \to 0$ as $\varepsilon \to 0$. From $d\chi_\varepsilon = \chi'_\varepsilon dr$ and $|dr|_g = 1$, it follows that $|d\chi_\varepsilon|_g^2 = |\chi'_\varepsilon dr|_g^2 = |\chi'_\varepsilon|^2$. Since $a \ge \frac{1}{n}$, we obtain

$$\begin{aligned} \int_M |d\chi_\varepsilon|_g^2 d\mu_g &= \int_0^l \int_N |\chi'_\varepsilon(r)|^2 f^n(r)\, dr d\mu_h \\ &\le C^n \operatorname{vol}(N,h) \int_\varepsilon^{\sqrt{\varepsilon}} |\chi'_\varepsilon|^2 r^{an} dr \quad (\text{by } f(r) \le Cr^a) \\ &= \frac{4C^n \operatorname{vol}(N,h)}{|\log \varepsilon|^2} \int_\varepsilon^{\sqrt{\varepsilon}} \left|\frac{1}{r}\right|^2 r^{an} dr \\ &= \frac{4C^n \operatorname{vol}(N,h)}{|\log \varepsilon|^2} \int_\varepsilon^{\sqrt{\varepsilon}} r^{an-2} dr \\ &= \frac{4C^n \operatorname{vol}(N,h)}{|\log \varepsilon|^2} \begin{cases} \dfrac{1}{an-1} \left[r^{an-1}\right]_\varepsilon^{\sqrt{\varepsilon}} & \text{if } an > 1, \\ \left[\log r\right]_\varepsilon^{\sqrt{\varepsilon}} & \text{if } an = 1 \end{cases} \\ &= 4C^n \operatorname{vol}(N,h) \begin{cases} \dfrac{1}{an-1} \cdot \dfrac{\varepsilon^{\frac{an-1}{2}} - \varepsilon^{an-1}}{|\log \varepsilon|^2} & \text{if } an > 1, \\ \dfrac{1}{2|\log \varepsilon|} & \text{if } an = 1 \end{cases} \\ &\longrightarrow 0 \quad (\text{as } \varepsilon \to 0). \end{aligned} \tag{9}$$

Therefore, from the Equations (8) and (9), we find that $\operatorname{Cap}(\partial_c M) = \operatorname{Cap}(\{x_0\}) = 0$. □

4. The Proof of Theorem 2

Proof of Theorem 2. Finally, we prove Theorem 2. We take an n-dimensional closed manifold (N^n, h) as the flat n-torus $(\mathbb{T}^n, h)$, where h is a flat metric on $\mathbb{T}^n$. We take the interval $I = (0, \pi)$ (i.e., $l = \pi$) and the warping function $f(r) := \sin^{1/n}(r)$, where $a := \frac{1}{n}$. Of course, this function $f(r)$ satisfies $f(r) > 0$ on $(0, \pi)$ and $f(+0) = f(-\pi) = 0$. Furthermore, there exists a positive constant $C > 0$ such that $C^{-1} r^a \le f(r) \le Cr^a$ on $(0, \pi)$.

Then, we consider the warped product Riemannian manifold $(M^{n+1}, g) = ((0, \pi) \times \mathbb{T}^n, dr^2 \oplus \sin^{2a}(r) h)$, which is homeomorphic to the regular set of the suspension $\Sigma(\mathbb{T}^n)$ of $\mathbb{T}^n$. This incomplete Riemannian manifold (M^{n+1}, g) is gluing two copies of the regular set $C_{\sin^a(r)}(\mathbb{T}^n)_{\text{reg}}$ along their boundaries:

$$(M^{n+1}, g) = C_{\sin^a(r)}(\mathbb{T}^n)_{\text{reg}} \cup_{\mathbb{T}^n} \Big(- C_{\sin^a(r)}(\mathbb{T}^n)_{\text{reg}} \Big),$$

where $-$ means the opposite orientation. By means of the partition of the unity, it is enough to show the properties (1) through (4) in Theorem 2 on the one side horn $C_{\sin^a(r)}(\mathbb{T}^n)_{\text{reg}} = ((0, \frac{\pi}{2}) \times \mathbb{T}^n, dr^2 \oplus \sin^{2a}(r)h)$.

Indeed,

(1) Since $f(r) = \sin^a(r)$ with $a = \frac{1}{n}$ and $\text{Ric}_h \equiv 0$, by Lemma 6, we have

- $\text{Ric}_g(\tilde{e}_0, \tilde{e}_0) = g(\text{Ric}_g(\tilde{e}_0), \tilde{e}_0) = na \left\{ 1 + (1-a)\dfrac{\cos^2(r)}{\sin^2(r)} \right\} \geq 1 > 0;$
- $\text{Ric}_g(\tilde{e}_i, \tilde{e}_i) = g(\text{Ric}_g(\tilde{e}_i), \tilde{e}_i) \geq a \left\{ 1 + (1-na)\dfrac{\cos^2(r)}{\sin^2(r)} \right\} = \dfrac{1}{n} > 0,$
  $(i = 1, \ldots, n).$

Hence, we see that the Ricci curvature of (M, g) satisfies $\text{Ric}_g \geq \dfrac{1}{n} =: K > 0$.

(2) Since $H^p(\mathbb{T}^n; \mathbb{R}) \neq 0$, by Lemmas 2 and 3, there exist non-trivial L^2 harmonic p-forms on (M, g) for all $1 \leq p \leq n-1$.

(3) In Lemma 5, since $a = \dfrac{1}{n}$, the range of p is $0 < p < n$. Hence, the L^2-Stokes theorem for p-forms with all $1 \leq p \leq n-1$ does not hold on (M, g).

(4) From Lemma 7, we see $\text{Cap}(\overline{M}_{\text{sing}}) = 0$.

□

5. Conclusions

A closed, more generally, complete Riemannian manifold with positive Ricci curvature satisfies the Bochner vanishing theorem. But, as we mentioned above, an incomplete Riemannian manifold does not satisfy a Bochner-type theorem in general. A key point is that the L^2-Stokes theorem does not hold. So, the author thinks that it would be important to study incomplete Riemannian manifolds where the L^2-Stokes theorem does not hold. Therefore, new phenomena might be discovered in geometry and analysis on manifolds with singularities.

Acknowledgments: The author is grateful to Jun Masamune for valuable discussion. The author is also grateful to the referees for helpful comments. The author is supported by the Grants-in-Aid for Scientific Research (C), Japan Society for the Promotion of Science, No. 16K05117.

Conflicts of Interest: The authors declare no conflict of interest.

References

1. Jost, J. Riemannian Geometry and Geometric Analysis. In *Universitext*, 6th ed.; Springer: Berlin, Germany, 2011.
2. Cheeger, J. On the Hodge theory of Riemannian pseudomanifolds. In *Proceedings of Symposia in Pure Mathematics*; AMS: Providence, RI, USA, 1980; Volume 36, pp. 91–146.
3. Cheeger, J. Spectral geometry of singular Riemannian spaces. *J. Differ. Geom.* **1983**, *18*, 575–657. [CrossRef]
4. Cheeger, J.; Goresky, M.; MacPherson, R. L^2-cohomology and intersection homology of singular algebraic varieties, Seminar on Differential Geometry. *Ann. Math. Stud.* **1982**, *102*, 303–340.
5. Bei, F. General perversities and L^2 de Rham and Hodge theorems for stratified pseudomanifolds. *Bull. Sci. Math.* **2014**, *138*, 2–40. [CrossRef]
6. Kirwan, F.; Woolf, J. *An Introduction to Intersection Homology Theory*, 2nd ed.; Chapman & Hall/CRC: London, UK, 2006.
7. Nagase, M. L^2-cohomology and intersection homology of stratified spaces. *Duke Math. J.* **1983**, *50*, 329–368. [CrossRef]

8. Youssin, B. L^p cohomology of cones and horns. *J. Differ. Geom.* **1994**, *39*, 559–603. [CrossRef]
9. Albin, P.; Leichtnam, E.; Mazzeo, R.; Piazza, P. Hodge theory on Cheeger spaces. *J. Reine Angew. Math.* **2018**, in press. doi:10.1515/crelle-2015-0095. [CrossRef]
10. Cheeger, J. A vanishing theorem for piecewise constant curvature spaces. In *Curvature and Topology of Riemannian Manifolds (Katata, 1985)*; Lect. Notes in Math.; Springer: Berlin, Germany, 1986; Volume 1201, pp. 33–40.
11. Dodziuk, J. Vanishing theorems for square-integrable harmonic forms. *Proc. Indian Acad. Sci. Math. Sci.* **1981**, *90*, 21–27. [CrossRef]
12. Gallot, S.; Meyer, D. Opérateur de courbure et laplacien des formes différentielles d'une variété riemannienne. *J. Math. Pures Appl.* **1975**, *54*, 259–284.
13. Petersen, P. Riemannian Geometry. In *GTM*, 3rd ed.; Springer: Berlin, Germany, 2016; Volume 171.
14. Brüning, J.; Lesch, M. Hilbert complexes. *J. Funct. Anal.* **1992**, *108*, 88–132. [CrossRef]
15. Grieser, D.; Lesch, M.L. On the L^2-Stokes theorem and Hodge theory for singular algebraic varieties. *Math. Nachr.* **2002**, *246–247*, 68–82. [CrossRef]
16. Behrens, S. The L^2 Stokes Theorem on Certain Incomplete Manifolds. Diploma Thesis, Univ. Bonn, Bonn, Germany, 2009.
17. Gaffney, M.P. The harmonic operator for exterior differential forms. *Proc. Natl. Acad. Sci. USA* **1951**, *37*, 48–50. [CrossRef] [PubMed]
18. Gaffney, M.P. A special Stokes' theorem for complete Riemannian manifolds. *Ann. Math.* **1954**, *60*, 140–145. [CrossRef]
19. Grigor'yan, A.; Masamune, J. Parabolicity and stochastic completeness of manifolds in terms of the Green formula. *J. Math. Pures Appl.* **2013**, *100*, 607–632. [CrossRef]
20. Brüning, J.; Lesch, M. Kähler-Hodge theory for conformal complex cones. *Geom. Funct. Anal.* **1993**, *3*, 439–473. [CrossRef]
21. Hunsicker, E.; Mazzeo, R. Harmonic forms on manifolds with edges. *Int. Math. Res. Not.* **2005**, *52*, 3229–3272. [CrossRef]
22. Besse, A. Einstein Manifolds. In *Ergebnisse der Mathematik und ihrer Grenzgebiete*; Band 10; Springer: Berlin, Germany, 1987.
23. Fukushima, M.; Ōshima, Y.; Takeda, M. *Dirichlet Forms and Symmetric Markov Processes, de Gruyter Studies in Math*; Walter de Gruyter: Berlin, Germany, 1994; Volume 19.

Σ mathematics

MDPI

Article

On Angles and Pseudo-Angles in Minkowskian Planes

Leopold Verstraelen

Department of Mathematics, Katholieke Universiteit Leuven, 3001 Leuven, Belgium; leopold.verstraelen@wis.kuleuven.be

Received: 1 March 2018; Accepted: 22 March 2018; Published: 3 April 2018

Abstract: The main purpose of the present paper is to well define Minkowskian angles and pseudo-angles between the two null directions and between a null direction and any non-null direction, respectively. Moreover, in a kind of way that will be tried to be made clear at the end of the paper, these new sorts of angles and pseudo-angles can similarly to the previously known angles be seen as (combinations of) Minkowskian lengths of arcs on a Minkowskian unit circle together with Minkowskian pseudo-lengths of parts of the straight null lines.

Keywords: Minkowski plane; Minkowskian length; Minkowskian angle; Minkowskian pseudo-angle

---

1. Introduction

In his 1908 lecture "Raum und Zeit" (cfr. Figure 1), Hermann Minkowski presented his indefinite geometry, which made possible the development of *Lorentzian geometry* and, more generally, of *pseudo-Euclidean geometry* and of *pseudo-Riemannian geometry*; (for references on these geometries, see e.g., [1–9] and the references in these books and chapters of books and articles).

XXXII.

Raum und Zeit.

(Vortrag, gehalten auf der 80. Naturforscher-Versammlung zu Köln am 21. Sept. 1908.)
(Physikalische Zeitschrift, 10. Jahrgang, 1909, S. 104—111 und Jahresbericht der Deutschen Mathematiker-Vereinigung, Bd. 18, S. 75—88; auch als Sonderabdruck erschienen, Leipzig, B. G. Teubner 1909.)*)

M. H.! Die Anschauungen über Raum und Zeit, die ich Ihnen entwickeln möchte, sind auf experimentell-physikalischem Boden erwachsen. Darin liegt ihre Stärke. Ihre Tendenz ist eine radikale. Von Stund an sollen Raum für sich und Zeit für sich völlig zu Schatten herabsinken, und nur noch eine Art Union der beiden soll Selbständigkeit bewahren.

*) Dem Vortrag war im Sonderabdruck ein Bildnis Minkowskis als Titelbild beigegeben. (Anm. d. Herausg.)

Figure 1. From Minkowski's "Raum und Zeit".

In the course of time, in *Minkowskian planes*, proper definitions have been given for *the angles between any two spacelike directions* and for *the angles between any two timelike directions* (which two cases geometrically are the same, of course) and for *the angle between any spacelike direction and any timelike direction*. A notion of such angles as equivalence classes under Minkowskian rotations, and their therefrom coming measure of angles has proper meaning only for two spacelike directions and equivalently for two timelike directions which direct to one and the same branch of a Minkowskian circle, or, still, to one and the same branch of a Euclidean orthogonal hyperbola with the two null directions of the Minkowskian plane as asymptotic directions, centered at the center of such rotations. And the measures of such angles then in fact are given by the Minkowskian lengths of the corresponding arcs on the concerned branches of this unit Minkowskian circle, in a perfect double analogy with the Euclidean angles between any two directions and their measure as central angles on a unit Euclidean circle. For the other above-mentioned Minkowskian angles then, the term angle essentially refers just to a measure of angle in some generalised above sense whereby trying to deal a bit cautiously with positive and negative arcs on Minkowskian circles. And further on, geometrically, the terms Minkowskian angles between any spacelike and any timelike directions—angles involving directions of one or both causal characters alike—will always refer to this common interpretation as central Minkowskian angles.

The main purpose of the present paper is to moreover well define *Minkowskian angles and pseudo-angles between the two null directions and between a null direction and any non-null direction, respectively*. Moreover, in a kind of way that will be tried to be made clear at the end of the paper, these new sorts of angles and pseudo-angles can similarly to the previously known angles be seen as (combinations of) *Minkowskian lengths of arcs on a Minkowskian unit circle together with Minkowskian pseudo-lengths of parts of the straight null lines*.

While all above central Minkowskian angles are invariant under Minkowskian rotations, the central Minkowskian pseudo-angles are not. However, in all cases—at least in the author's opinion—all of these Minkowskian angles and pseudo-angles may turn out not to be without some uses in geometry and in applications of geometry. In this respect, for instance: (i) they may give some geometrical interpretations of the so far merely algebraical or analytical conditions that did occur in some articles on geometry and its applications; (ii) they may lead to extensions of several theories about Riemannian submanifolds in which Euclidean angles play a role to corresponding theories in pseudo-Riemannian geometry; (iii) they may be studied by working out a corresponding trigonometry; (iv) they may be extended to suitable notions for angles and pseudo-angles between higher-dimensional linear subspaces of various causal natures in general pseudo-Euclidean spaces; etcetera.

For related literature on geometry and its applications, see, e.g., also [10–26].

The point of departure of this paper is the definition of *the pseudo-angles or "angles" between any two vectors in a Minkowskian plane as given by Garry Helzer in his relativistic version of the formula of Gauss-Bonnet* [27].

In the present paper, like in several of his other recent papers, when it seems to the author of real importance for a better understanding of the text, he included a number of handmade figures. In his experience, so much more than artificially made illustrations, such figures do essentially contribute to the readability of the paper. This is very related to the real value of the drawings made on blackboards during proper lectures on mathematics and on the exact sciences. The author is very grateful for the editors of the journal Mathematics having been so kind to include the scans of ten handmade figures in the present paper.

2. The Pseudo-Angles of Helzer

Let E^2_1 be *the Minkowskian plane* (R^2, g) fixed by the $(+,-)$ metric

$$g(\vec{v}, \vec{w}) = v_1 w_1 - v_2 w_2 \tag{1}$$

on the standard two-dimensional real vector space R^2, whereby $\vec{v} = (v_1, v_2)$ and $\vec{w} = (w_1, w_2)$ here denote arbitrary vectors in R^2 expanded with respect to *the standard oriented orthonormal basis* $B = \{\vec{e_1}, \vec{e_2}\}$. Next, let φ_B be *the real valued function which is defined on the set S of the unit vectors and of the null vectors in* E^2_1, that is, on the set of the vectors $\vec{z} = z_1\vec{e_1} + z_2\vec{e_2}$ for which $z_1^2 - z_2^2 = \pm 1$ (i.e., *on the two Euclidean unit orthogonal hyperbola's* $H_1 : z_1^2 - z_2^2 = 1$ and $H_2 : z_1^2 - z_2^2 = -1$) and on the vectors $\vec{z} = z_1\vec{e_1} + z_2\vec{e_2} \neq \vec{o}$ for which $z_1^2 - z_2^2 = 0$ (i.e., *on the first and second diagonals or Euclidean bisectrices* $D_1 : z_1 - z_2 = 0$ *and* $D_2 : z_1 + z_2 = 0$*, "minus" the origin O*), by

$$\varphi_B(\vec{z}) = \begin{cases} \ln|z_1 + z_2|, & when \quad z_1 + z_2 \neq 0, \\ -\ln|z_1 - z_2|, & when \quad z_1 + z_2 = 0. \end{cases} \tag{2}$$

And finally, let ψ_B be *the real valued function which is defined on the pairs of vectors from S*, say $\vec{v}$ and $\vec{w}$, by

$$\psi_B(\vec{v}, \vec{w}) = \varphi_B(\vec{w}) - \varphi_B(\vec{v}). \tag{3}$$

when similarly defining functions $\varphi_{B'}$ and $\psi_{B'}$ corresponding to *any other ordered orthonormal basis* B' of E_1^2, then $\psi_{B'} = \psi_B$ or $\psi_{B'} = -\psi_B$, *depending on* B' *and* B *having the same or opposite orientations*, respectively; (as a kind of intermediate step in this having $\varphi_{B'}(\vec{z}) = \varphi_B(\vec{z}) + \varphi_{B'}(\vec{e}_1)$ and $\varphi_{B'}(\vec{z}) = -\varphi_B(\vec{z}) + \varphi_{B'}(\vec{e}_1)$, respectively).

Therefore, the following makes good sense indeed: *the oriented pseudo-angle* $\psi(\vec{v}, \vec{w})$ *of Helzer* [27] *from* $\vec{v}$ *to* $\vec{w}$, $(\vec{v}, \vec{w} \in S)$, *is defined* by

$$\psi(\vec{v}, \vec{w}) = \psi_B(\vec{v}, \vec{w}). \tag{4}$$

According to this definition, clearly

$$\psi(\vec{v}, \vec{w}) = \psi(-\vec{v}, \vec{w}) = \psi(\vec{v}, -\vec{w}) = \psi(-\vec{v}, -\vec{w}). \tag{5}$$

And for any number k of unit or null vectors $\vec{v_1}, \vec{v_2}, \ldots, \vec{v_k}$ in E_1^2, one has

$$\psi(\vec{v_1}, \vec{v_2}) + \psi(\vec{v_2}, \vec{v_3}) + \cdots + \psi(\vec{v}_{k-1}, \vec{v_k}) + \psi(\vec{v_k}, \vec{v_1}) = 0. \tag{6}$$

3. The Minkowskian Angles between Spacelike and Timelike Directions

The following result shows that *for unit spacelike or timelike vectors* $\vec{v}$ *and* $\vec{w}$ *in a Minkowskian plane* E_1^2 *the oriented pseudo-angle* $\psi(\vec{v}, \vec{w})$ *of Helzer is equal to what O'Neill in* [5] *called the oriented Lorentz angle between two spacelike unit vectors* $\vec{v}$ *and* $\vec{w}$, *or is equal to what Birman and Nomizu in* [28,29] *simply called the oriented angle between two timelike unit vectors* $\vec{v}$ *and* $\vec{w}$, *or is equal to what in* [30,31] *was called the oriented hyperbolic angle between a spacelike unit vector* $\vec{v}$ *and a timelike unit vector* $\vec{w}$, *depending on the causal characters of* $\vec{v}$ *and* $\vec{w}$.

Before giving the formulation and a proof of this result, I would like to make the following proposal concerning terminology: *let us use the term "Minkowskian angles"* when dealing with the above kind of angles between unit vectors, and also between their directions in a Minkowskian plane, (rather than just "angles", since angles as such are commonly used for the common angles of Euclidean geometry, and rather than *"Lorentzian angles"*, since also on Lorentzian surfaces the angles are essentially defined in the tangent planes to such surfaces and these are Minkowskian planes, and also rather than *"hyperbolic angles"*, which seem better to be reserved for use in the geometry of Lobachevsky-Bolyai; see also Section 7 concerning this matter).

Theorem 1. *Let* $\vec{v}$ *and* $\vec{w}$ *be unit vectors in a Minkowskian plane* E_1^2 *and let* $\psi(\vec{v}, \vec{w})$ *be the oriented pseudo-angle of Helzer from* $\vec{v}$ *to* $\vec{w}$. *Then, when* (v_1, v_2) *and* (w_1, w_2) *are the co-ordinates of* $\vec{v}$ *and* $\vec{w}$ *with respect to the standard basis B in* E_1^2 *and when D is the Euclidean reflection in the first diagonal of B, in terms of the hyperbolic functions cosh and sinh, this pseudo-angle* ψ *is related to the Minkowskian metric* g *in the following way:*
(i) if $\vec{v}$ *and* $\vec{w}$ *are both spacelike,*

$$\cosh\psi(\vec{v}, \vec{w}) = \begin{cases} -g(\vec{v}, \vec{w}) & when \quad sgn\, v_1 \neq sgn\, w_1 \\ g(\vec{v}, \vec{w}) & when \quad sgn\, v_1 = sgn\, w_1, \end{cases} \tag{7}$$

$$\sinh\psi(\vec{v}, \vec{w}) = \begin{cases} -g(\vec{v}, D\vec{w}) & when \quad sgn\, v_1 \neq sgn\, w_1 \\ g(\vec{v}, D\vec{w}) & when \quad sgn\, v_1 = sgn\, w_1; \end{cases} \tag{8}$$

(ii) if $\overrightarrow{v}$ and $\overrightarrow{w}$ are both timelike,

$$\cosh\psi(\overrightarrow{v},\overrightarrow{w}) = \begin{cases} g(\overrightarrow{v},\overrightarrow{w}) & when \quad sgn\, v_2 \neq sgn\, w_2 \\ -g(\overrightarrow{v},\overrightarrow{w}) & when \quad sgn\, v_2 = sgn\, w_2, \end{cases} \tag{9}$$

$$\sinh\psi(\overrightarrow{v},\overrightarrow{w}) = \begin{cases} g(\overrightarrow{v},D\overrightarrow{w}) & when \quad sgn\, v_2 \neq sgn\, w_2, \\ -g(\overrightarrow{v},D\overrightarrow{w}) & when \quad sgn\, v_2 = sgn\, w_2; \end{cases} \tag{10}$$

(iii) if $\overrightarrow{v}$ is spacelike and $\overrightarrow{w}$ is timelike,

$$\cosh\psi(\overrightarrow{v},\overrightarrow{w}) = \begin{cases} -g(\overrightarrow{v},D\overrightarrow{w}) & when \quad sgn\, v_1 \neq sgn\, w_2 \\ g(\overrightarrow{v},D\overrightarrow{w}) & when \quad sgn\, v_1 = sgn\, w_2, \end{cases} \tag{11}$$

$$\sinh\psi(\overrightarrow{v},\overrightarrow{w}) = \begin{cases} -g(\overrightarrow{v},\overrightarrow{w}) & when \quad sgn\, v_1 \neq sgn\, w_2 \\ g(\overrightarrow{v},\overrightarrow{w}) & when \quad sgn\, v_1 = sgn\, w_2. \end{cases} \tag{12}$$

Proof. For any pair of *unit vectors* $\overrightarrow{v} = (v_1, v_2)$ and $\overrightarrow{w} = (w_1, w_2)$ according to (1)–(4),

$$\begin{aligned} \psi(\overrightarrow{v},\overrightarrow{w}) &= \ln|w_1+w_2| - \ln|v_1+v_2| \\ &= \ln|\frac{w_1+w_2}{v_1+v_2}| \\ &= \ln|\frac{(w_1+w_2)(v_1-v_2)}{v_1^2-v_2^2}| \\ &= \ln|(w_1+w_2)(v_1-v_2)| \\ &= \ln|v_1w_1 - v_2w_2 + v_1w_2 - v_2w_1|, \end{aligned}$$

and, so, since $D\overrightarrow{w} = D(w_1, w_2) = (w_2, w_1)$,

$$\psi(\overrightarrow{v},\overrightarrow{w}) = \ln|g(\overrightarrow{v},\overrightarrow{w}) + g(\overrightarrow{v},D\overrightarrow{w})|. \tag{13}$$

Hence, by the very definitions of the functions cosinushyperbolicus and sinushyperbolicus,

$$\cosh\psi(\overrightarrow{v},\overrightarrow{w}) = \frac{[g(\overrightarrow{v},\overrightarrow{w}) + g(\overrightarrow{v},D\overrightarrow{w})]^2 + 1}{2|g(\overrightarrow{v},\overrightarrow{w}) + g(\overrightarrow{v},D\overrightarrow{w})|} \tag{14}$$

and

$$\sinh\psi(\overrightarrow{v},\overrightarrow{w}) = \frac{[g(\overrightarrow{v},\overrightarrow{w}) + g(\overrightarrow{v},D\overrightarrow{w})]^2 - 1}{2|g(\overrightarrow{v},\overrightarrow{w}) + g(\overrightarrow{v},D\overrightarrow{w})|}. \tag{15}$$

In the cases (i) and (ii), i.e., if $\overrightarrow{v}$ and $\overrightarrow{w}$ either are *both spacelike* (i) or are *both timelike* (ii),

$$g(\overrightarrow{v},\overrightarrow{w})^2 - g(\overrightarrow{v},D\overrightarrow{w})^2 = 1, \tag{16}$$

which combined with (14) and (15), yields

$$\cosh\psi(\overrightarrow{v},\overrightarrow{w}) = \epsilon\, g(\overrightarrow{v},\overrightarrow{w}) \tag{17}$$

and

$$\sinh\psi(\overrightarrow{v},\overrightarrow{w}) = \epsilon\, g(\overrightarrow{v},D\overrightarrow{w}), \tag{18}$$

whereby $\epsilon = sgn[g(\overrightarrow{v},\overrightarrow{w}) + g(\overrightarrow{v},D\overrightarrow{w})]$. In addition, then formulae (7) and (8) and formulae (9) and (10) follow from formulae (17) and (18) since, *when $\overrightarrow{v}$ and $\overrightarrow{w}$ are both spacelike*, $\epsilon = 1$ when

$sgn\, v_1 = sgn\, w_1$ and $\epsilon = -1$ when $sgn\, v_1 \neq sgn\, w_1$, and *when* $\overrightarrow{v}$ *and* $\overrightarrow{w}$ *are both timelike*, $\epsilon = 1$ when $sgn\, v_2 \neq sgn\, w_2$ and $\epsilon = -1$ when $sgn\, v_2 = sgn\, w_2$.

Finally, in case (iii), i.e., if $\overrightarrow{v}$ and $\overrightarrow{w}$ do have *different causal characters*, say *if* $\overrightarrow{v}$ *is spacelike and* $\overrightarrow{w}$ *is timelike*,

$$g(\overrightarrow{v}, \overrightarrow{w})^2 - g(\overrightarrow{v}, D\overrightarrow{w})^2 = -1, \tag{19}$$

which, combined with (14) and (15), now yields

$$\cosh \psi(\overrightarrow{v}, \overrightarrow{w}) = \epsilon\, g(\overrightarrow{v}, D\overrightarrow{w}) \tag{20}$$

and

$$\sinh \psi(\overrightarrow{v}, \overrightarrow{w}) = \epsilon\, g(\overrightarrow{v}, \overrightarrow{w}). \tag{21}$$

In addition, then formulae (11) and (12) follow from formulae (20) and (21) since, *for a spacelike unit vector* $\overrightarrow{v}$ *and a timelike unit vector* $\overrightarrow{w}$, $\epsilon = -1$ when $sgn\, v_1 \neq sgn\, w_2$ and $\epsilon = 1$ when $sgn\, v_1 = sgn\, w_2$. □

From here on, we agree to systematically use the *notation* $\theta(\overrightarrow{v}, \overrightarrow{w})$ *for the oriented Minkowskian angles between unit vectors* $\overrightarrow{v}$ *and* $\overrightarrow{w}$ *for any causal characters*, (rather than the former $\psi(\overrightarrow{v}, \overrightarrow{w})$, keeping on the use of ψ though for the pseudo-angles of Helzer in general, cfr. definition (3)).

In the Minkowskian geometry on a plane, unit vectors $\overrightarrow{v}$ and $\overrightarrow{w}$ for which $g(\overrightarrow{v}, \overrightarrow{w}) = 0$, apart from pairs of vectors $\pm\overrightarrow{e_1}$ and $\pm\overrightarrow{e_2}$, are not at all orthogonal or perpendicular to each other in accordance with our common visual senses, or, still, in accordance with the Euclidean geometry on this plane. However, such unit vectors in a Minkowskian plane, i.e., *unit vectors in* E^2_1 *with vanishing Minkowskian scalar product*, nevertheless, conventionally *often remain said to be mutually orthogonal*. All in all, this terminology may not be so recommendable (but, unfortunately, it is to be expected that this terminology will continue to be used, like; for instance, one has been going on to speak of "the orthogonal group" when speaking of "the orthonormal group"...). Actually, such vectors are each other's Euclidean reflections in the first or second diagonals $D = D_1$ and D_2 of the standard orthonormal basis $B = \{\overrightarrow{e_1}, \overrightarrow{e_2}\}$, or, put otherwise, *such vectors* are pairs of vectors lying on the Euclidean orthogonal hyperbola's $H : u_1^2 - u_2^2 = \pm 1$ with Euclidean unit axes and *which are bisected either by the first or second diagonals or bisectrices* D_1 *and* D_2 (cfr. Figure 2). It could be observed here in passing, and it will become more clear later on, that the just used expressions that refer to bisecting, however, do enjoy their proper meanings in the sense of the angles in the geometries of Euclid and of Minkowski alike. In any case, based on Theorem 1, one has the following.

Corollary 1. *While in a Minkowskian plane any two unit vectors with the same causal character can never be mutually orthogonal, a timelike and a spacelike unit vector are mutually orthogonal if and only if their oriented Minkowskian angle is zero.*

For two arbitrary (non-null) vectors $\overrightarrow{a} = (a_1, a_2)$ *and* $\overrightarrow{b} = (b_1, b_2)$ of Minkowskian lengths $||\overrightarrow{a}|| = |g(\overrightarrow{a}, \overrightarrow{a})|^{\frac{1}{2}} (\neq 0)$ and $||\overrightarrow{b}|| = |g(\overrightarrow{b}, \overrightarrow{b})|^{\frac{1}{2}} (\neq 0)$, *from* $\overrightarrow{a}$ *to* $\overrightarrow{b}$ *the oriented Minkowskian angle* $\theta(\overrightarrow{a}, \overrightarrow{b})$ *is defined as the oriented Minkowskian angle of their normalised corresponding unit vectors* $\overrightarrow{v} = \overrightarrow{a}/||\overrightarrow{a}||$ and $\overrightarrow{w} = \overrightarrow{b}/||\overrightarrow{b}||$; $\theta(\overrightarrow{a}, \overrightarrow{b}) = \theta(\overrightarrow{v}, \overrightarrow{w})$, (cfr. Figure 3). Thus, according to (13), in terms of the Minkowskian scalar product:

$$\theta(\overrightarrow{a}, \overrightarrow{b}) = \ln \left| \frac{g(\overrightarrow{a}, \overrightarrow{b}) + g(\overrightarrow{a}, D\overrightarrow{b})}{||\overrightarrow{a}|| \cdot ||\overrightarrow{b}||} \right|. \tag{22}$$

Further, based on relations (1)–(6), for all pairs of arbitrary non-null vectors $\overrightarrow{a}$ and $\overrightarrow{b}$ we recover *the following formulae which relate these oriented Minkowskian angles* $\theta(\overrightarrow{a}, \overrightarrow{b})$ *to the Minkowskian metric by means of the hyperbolic functions*, (cfr. [28]).

$$g\left(\vec{v}, D_1\vec{v}\right) = g\left(\vec{v}, D_2\vec{v}\right) = 0$$
$$g\left(\vec{e_1}, \vec{e_2}\right) = g\left(\vec{e_1}, -\vec{e_2}\right) = 0$$

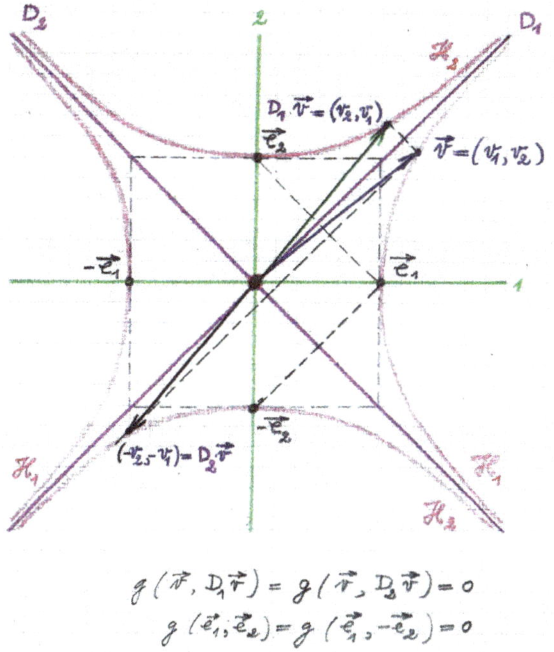

Figure 2. "Orthogonality".

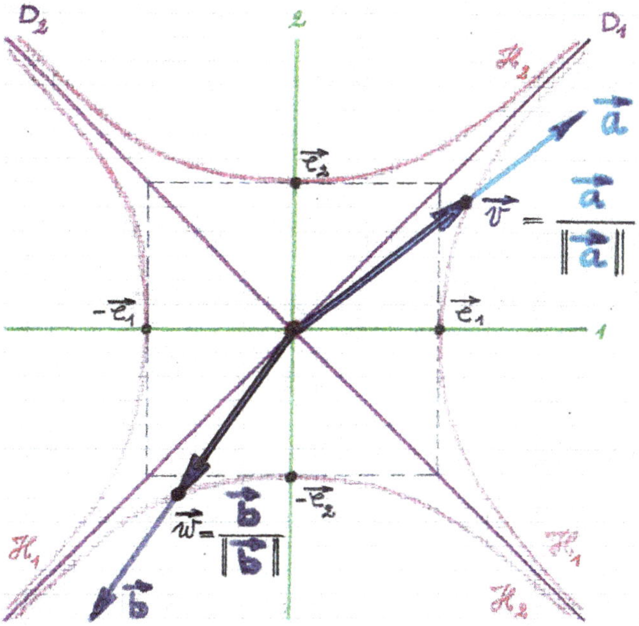

Figure 3. Angles between non-null vectors.

Theorem 2. *(i) If $\overrightarrow{a}$ and $\overrightarrow{b}$ are both spacelike,*

$$\cosh\theta(\overrightarrow{a},\overrightarrow{b}) = \begin{cases} -\dfrac{g(\overrightarrow{a},\overrightarrow{b})}{||\overrightarrow{a}||\cdot||\overrightarrow{b}||} & when \quad sgn\, a_1 \neq sgn\, b_1 \\ \dfrac{g(\overrightarrow{a},\overrightarrow{b})}{||\overrightarrow{a}||\cdot||\overrightarrow{b}||} & when \quad sgn\, a_1 = sgn\, b_1, \end{cases} \tag{23}$$

$$\sinh\theta(\overrightarrow{a},\overrightarrow{b}) = \begin{cases} -\dfrac{g(\overrightarrow{a},D\overrightarrow{b})}{||\overrightarrow{a}||\cdot||\overrightarrow{b}||} & when \quad sgn\, a_1 \neq sgn\, b_1 \\ \dfrac{g(\overrightarrow{a},D\overrightarrow{b})}{||\overrightarrow{a}||\cdot||\overrightarrow{b}||} & when \quad sgn\, a_1 = sgn\, b_1; \end{cases} \tag{24}$$

(ii) if $\overrightarrow{a}$ and $\overrightarrow{b}$ are both timelike,

$$\cosh\theta(\overrightarrow{a},\overrightarrow{b}) = \begin{cases} \dfrac{g(\overrightarrow{a},\overrightarrow{b})}{||\overrightarrow{a}||\cdot||\overrightarrow{b}||} & when \quad sgn\, a_2 \neq sgn\, b_2 \\ -\dfrac{g(\overrightarrow{a},\overrightarrow{b})}{||\overrightarrow{a}||\cdot||\overrightarrow{b}||} & when \quad sgn\, a_2 = sgn\, b_2, \end{cases} \tag{25}$$

$$\sinh\theta(\overrightarrow{a},\overrightarrow{b}) = \begin{cases} \dfrac{g(\overrightarrow{a},D\overrightarrow{b})}{||\overrightarrow{a}||\cdot||\overrightarrow{b}||} & when \quad sgn\, a_2 \neq sgn\, b_2 \\ -\dfrac{g(\overrightarrow{a},D\overrightarrow{b})}{||\overrightarrow{a}||\cdot||\overrightarrow{b}||} & when \quad sgn\, a_2 = sgn\, b_2; \end{cases} \tag{26}$$

(iii) if $\overrightarrow{a}$ is spacelike and $\overrightarrow{b}$ is timelike,

$$\cosh\theta(\overrightarrow{a},\overrightarrow{b}) = \begin{cases} -\dfrac{g(\overrightarrow{a},D\overrightarrow{b})}{||\overrightarrow{a}||\cdot||\overrightarrow{b}||} & when \quad sgn\, a_1 \neq sgn\, b_2 \\ \dfrac{g(\overrightarrow{a},D\overrightarrow{b})}{||\overrightarrow{a}||\cdot||\overrightarrow{b}||} & when \quad sgn\, a_1 = sgn\, b_2, \end{cases} \tag{27}$$

$$\sinh\theta(\overrightarrow{a},\overrightarrow{b}) = \begin{cases} -\dfrac{g(\overrightarrow{a},\overrightarrow{b})}{||\overrightarrow{a}||\cdot||\overrightarrow{b}||} & when \quad sgn\, a_1 \neq sgn\, b_2 \\ \dfrac{g(\overrightarrow{a},\overrightarrow{b})}{||\overrightarrow{a}||\cdot||\overrightarrow{b}||} & when \quad sgn\, a_1 = sgn\, b_2. \end{cases} \tag{28}$$

Based on the definitions given above for the oriented Minkowskian angles between any two spacelike or timelike vectors, and also in view of (5), *the oriented Minkowskian angle $\theta(L_1, L_2)$ between any two non-null directions* or *between any non-null lines L_1 and L_2 passing through the origin of a Minkowskian plane E_1^2* may be well defined as the oriented Minkowskian angle $\theta(\overrightarrow{l_1},\overrightarrow{l_2})$ between a unit vector $\overrightarrow{l_1}$ on the line L_1 and a unit vector $\overrightarrow{l_2}$ on the line L_2; $\theta(L_1, L_2) = \theta(\overrightarrow{l_1},\overrightarrow{l_2})$.

As a kind of transition to the definition of Minkowskian angles involving one or two null vectors and also in a way continuing the former comment on perpendicular vectors in a Minkowskian plane, now, (hereby somewhat following O'Neill [5], p. 48, Figure 3), one may *visualise*, for instance, *the following pair of vectors that are mutually orthogonal in E_1^2: $\{(n,m), D(n,m) = (m,n)|n \in R_0^+, m \in]0,n[\}$. A null vector like $\overrightarrow{v} = (n,n)$ may thus be seen* to originate *as the limit of the pair of the mutually orthogonal vectors formed by the spacelike vector (n,m) and the timelike vector (m,n) for m going to n: $(n,n) = lim_{m\to n}(n,m) = lim_{m\to n}(m,n)$,* this limit thus yielding *a non-trivial vector $\overrightarrow{n}$ that is* *perpendicular to itself*; (cfr. Figure 4). And of course, similarly one may think of the null vectors of the second diagonal too as *non-trivial auto-orthogonal vectors*.

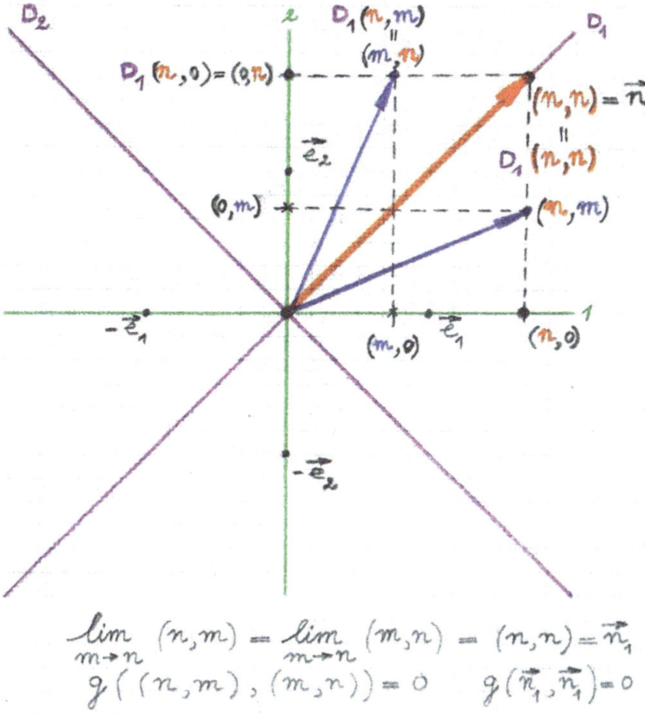

Figure 4. Auto-orthogonal vectors.

4. The Minkowskian Pseudo-Angles between A Null Direction and Any Spacelike or Timelike Direction

Any *spacelike or timelike vector* $\overrightarrow{a} = (a_1, a_2)$ has a well determined *Minkowskian length or norm* $||\overrightarrow{a}|| = |a_1^2 - a_2^2|^{\frac{1}{2}}$ *which is essentially non-zero* and thus *such a vector can be normalised to the corresponding unit spacelike or timelike vector* $\overrightarrow{v} = \overrightarrow{a}/||\overrightarrow{a}||$. For *null vectors* $\overrightarrow{n_1} = (n, n)$ *of* D_1 *and* $\overrightarrow{n_2} = (-n, n)$ *of* D_2, $(n \in R_0)$, this kind of normalisation of course is not possible, since actually $||\overrightarrow{n_1}||^2 = ||\overrightarrow{n_2}||^2 = |n^2 - n^2| = 0$. Then, for null vectors in a Minkowskian plane choosing as a way of standardisation *the individual normalisation of their two components*, from now on, we propose to consider, respectively $\overrightarrow{d_1} = (1, 1) = \overrightarrow{n_1}/|n|$ in case $n > 0$ and $-\overrightarrow{d_1} = (-1, -1) = \overrightarrow{n_1}/|n|$ in case $n < 0$, and $\overrightarrow{d_2} = (-1, 1) = \overrightarrow{n_2}/|n|$ in case $n > 0$ and $-\overrightarrow{d_2} = (1, -1) = \overrightarrow{n_1}/|n|$ in case $n < 0$, as *the normalised null vectors corresponding to given null vectors* $\overrightarrow{n_1}$ *and* $\overrightarrow{n_2}$; (cfr. Figures 5 and 6). And while *for spacelike and timelike vectors* $\overrightarrow{a} = (a_1, a_2)$ *their norm equals their length* $||\overrightarrow{a}||$, *for null vectors* $\overrightarrow{n_1} = (n, n)$ *and* $\overrightarrow{n_2} = (-n, n)$, $n \in R_0$, *their lengths* $||\overrightarrow{n_1}||$ *and* $||\overrightarrow{n_2}||$ *being zero, we propose to define their pseudo-norms* $|\overrightarrow{n_1}|$ *and* $|\overrightarrow{n_2}|$ *to be equal to the absolute value* $|n| = |-n| \neq 0$ *of their co-ordinates*: $|\overrightarrow{n_1}| = |\overrightarrow{n_2}| = |n|$. Thus, we have the normalisations $\overrightarrow{d_i} = \pm\overrightarrow{n_i}/|\overrightarrow{n_i}|$, $+$ or $-$ depending on n being positive or negative, respectively.

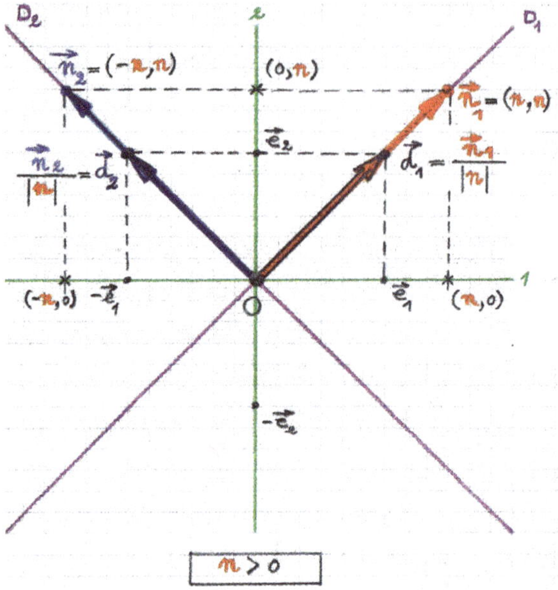

Figure 5. Two normalised null vectors.

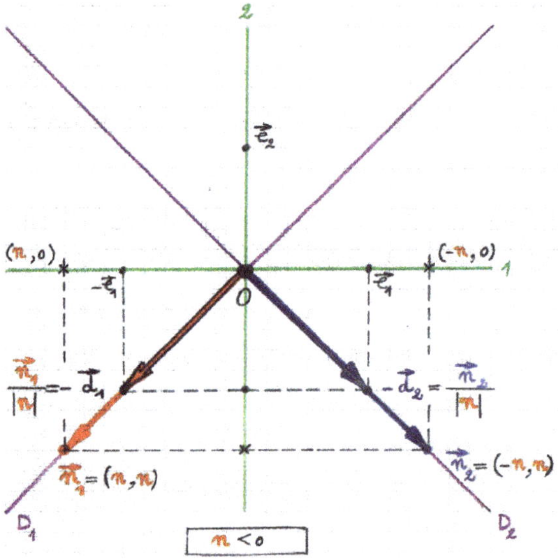

Figure 6. The two other normalised null vectors.

The oriented Minkowskian pseudo-angles θ *between unit spacelike or timelike vectors* $\overrightarrow{u} = (u_1, u_2)$, $(u_1^2 - u_2^2 = \pm 1)$, *and normalised null vectors* $\pm\overrightarrow{d_1}$ *or* $\pm\overrightarrow{d_2}$ *are defined to be given by their pseudo-angles* ψ *of Helzer.* In order to establish the formulae for these angles in terms of the hyperbolic functions, next the angles $\theta(\overrightarrow{u}, \overrightarrow{d_1})$ and $\theta(\overrightarrow{d_2}, \overrightarrow{u})$ will be dealt with explicitly; (thereby, in view of (5) and (6), essentially all possibilities are taken care of). *The oriented Minkowskian pseudo-angle* $\theta(\overrightarrow{u}, \overrightarrow{d_1})$ *from* $\overrightarrow{u}$ *to* $\overrightarrow{d_1}$ *is defined to be given by the pseudo-angle* $\psi(\overrightarrow{u}, \overrightarrow{d_1})$ *from* $\overrightarrow{u}$ *to* $\overrightarrow{d_1}$ such that according to (1)–(4),

$$\begin{aligned} \theta(\overrightarrow{u}, \overrightarrow{d_1}) &= \ln 2 - \ln |u_1 + u_2| \\ &= \ln |\tfrac{2}{u_1+u_2}| \\ &= \ln |\tfrac{2(u_1-u_2)}{u_1^2-u_2^2}| \\ &= \ln |2(u_1 - u_2)| \\ &= \ln |2g(\overrightarrow{u}, \overrightarrow{d_1})| \\ (&= \ln |g(\overrightarrow{u}, \overrightarrow{d_1}) + g(\overrightarrow{u}, D\overrightarrow{d_1})|). \end{aligned} \tag{29}$$

And, *the oriented Minkowskian pseudo-angle* $\theta(\overrightarrow{d_2}, \overrightarrow{u})$ *from* $\overrightarrow{d_2}$ *to* $\overrightarrow{u}$ *is defined to be given by the pseudo-angle* $\psi(\overrightarrow{d_2}, \overrightarrow{u})$ *from* $\overrightarrow{d_2}$ *to* $\overrightarrow{u}$ such that, according to (1)–(4),

$$\begin{aligned} \theta(\overrightarrow{d_2}, \overrightarrow{u}) &= \ln |u_1 + u_2| + \ln 2 \\ &= \ln |2(u_1 + u_2)| \\ &= \ln |2g(\overrightarrow{d_2}, \overrightarrow{u})| \\ (&= \ln |g(\overrightarrow{d_2}, \overrightarrow{u}) + g(\overrightarrow{d_2}, D\overrightarrow{u})|). \end{aligned} \tag{30}$$

Hence, in analogy with Theorem 1, from (29) and (30), one has the following.

Theorem 3. *Let* $\theta(\overrightarrow{u}, \overrightarrow{d_1})$ *and* $\theta(\overrightarrow{d_2}, \overrightarrow{u})$ *be the oriented Minkowskian pseudo-angles from a unit vector* $\overrightarrow{u}$ *to the normalised null vector* $\overrightarrow{d_1}$ *and from the normalised null vector* $\overrightarrow{d_2}$ *to a unit vector* $\overrightarrow{u}$, *respectively. Then,*

$$\cosh\theta(\overrightarrow{u}, \overrightarrow{d_1}) = \begin{cases} \dfrac{1+4g(\overrightarrow{u}, \overrightarrow{d_1})^2}{4g(\overrightarrow{u}, \overrightarrow{d_1})} & \textit{when} \quad u_1 - u_2 > 0 \\ -\dfrac{1+4g(\overrightarrow{u}, \overrightarrow{d_1})^2}{4g(\overrightarrow{u}, \overrightarrow{d_1})} & \textit{when} \quad u_1 - u_2 < 0, \end{cases} \tag{31}$$

$$\sinh\theta(\overrightarrow{u}, \overrightarrow{d_1}) = \begin{cases} \dfrac{1-4g(\overrightarrow{u}, \overrightarrow{d_1})^2}{4g(\overrightarrow{u}, \overrightarrow{d_1})} & \textit{when} \quad u_1 - u_2 < 0 \\ -\dfrac{1-4g(\overrightarrow{u}, \overrightarrow{d_1})^2}{4g(\overrightarrow{u}, \overrightarrow{d_1})} & \textit{when} \quad u_1 - u_2 > 0; \end{cases} \tag{32}$$

$$\cosh\theta(\overrightarrow{d_2}, \overrightarrow{u}) = \begin{cases} \dfrac{1+4g(\overrightarrow{d_2}, \overrightarrow{u})^2}{4g(\overrightarrow{d_2}, \overrightarrow{u})} & \textit{when} \quad u_1 + u_2 > 0 \\ -\dfrac{1+4g(\overrightarrow{d_2}, \overrightarrow{u})^2}{4g(\overrightarrow{d_2}, \overrightarrow{u})} & \textit{when} \quad u_1 + u_2 < 0, \end{cases} \tag{33}$$

$$\sinh\theta(\overrightarrow{d_2}, \overrightarrow{u}) = \begin{cases} \dfrac{1-4g(\overrightarrow{d_2}, \overrightarrow{u})^2}{4g(\overrightarrow{d_2}, \overrightarrow{u})} & \textit{when} \quad u_1 + u_2 < 0 \\ -\dfrac{1-4g(\overrightarrow{d_2}, \overrightarrow{u})^2}{4g(\overrightarrow{d_2}, \overrightarrow{u})} & \textit{when} \quad u_1 + u_2 > 0. \end{cases} \tag{34}$$

In connection with a general comment made in the Introduction concerning potential applications of the contents of this paper in semi-Riemannian geometry, based on Theorems 1 and 3, it may be good to explicitly formulate the following.

Corollary 2. *For any two different normalised spacelike, timelike or null vector fields, their Minkowskian scalar product is constant if and only if their Minkowskian angle or pseudo-angle is constant.*

For any spacelike or timelike vector $\overrightarrow{a}$ of Minkowskian length $||\overrightarrow{a}||$, $\overrightarrow{u} = \overrightarrow{a}/||\overrightarrow{a}||$ is the corresponding normalised spacelike or timelike vector, and for any null vector $\overrightarrow{n_1} = (n,n)$ or $\overrightarrow{n}_2 = (-n,n)$, $(n \in R_0)$, $\pm\overrightarrow{d_1} = \overrightarrow{n_1}/|\overrightarrow{n_1}|$ or $\pm\overrightarrow{d_2} = \overrightarrow{n_2}/|\overrightarrow{n_2}|$, (+ when $n > 0$ and $-$ when $n < 0$) is the corresponding normalised null vector, and *the oriented Minkowskian pseudo-angles between such vectors* $\overrightarrow{a}$ *and* $\overrightarrow{n_i}$, $(i = 1,2)$, *are defined as the oriented Minkowskian pseudo-angles between their corresponding normalised vectors* $\overrightarrow{u}$ *and* $\pm\overrightarrow{d_i}$; $\theta(\overrightarrow{a}, \overrightarrow{n_i}) = \theta(\overrightarrow{u}, \pm\overrightarrow{d_i})$. In addition, following this definition, the above expressions (31)–(34) may readily be adapted to corresponding formulae involving the Minkowskian scalar products $g(\overrightarrow{a}, \overrightarrow{n_i})$ as follows.

Theorem 4. *Let* $\theta(\overrightarrow{a}, \overrightarrow{n_1})$ *and* $\theta(\overrightarrow{n_2}, \overrightarrow{a})$ *be the oriented Minkowskian pseudo-angles from a non-null vector* $\overrightarrow{a} = (a_1, a_2)$ *of arbitrary length* $||\overrightarrow{a}|| \neq 0$ *to an arbitrary null vector* $\overrightarrow{n_1} = (n,n)$, $n \in R_0$ *and from an arbitrary null vector* $\overrightarrow{n_2} = (-n,n)$ *to a non-null vector* $\overrightarrow{a}$, *respectively. Then,*

$$\cosh\theta(\overrightarrow{a}, \overrightarrow{n_1}) = \begin{cases} \dfrac{|n|^2||\overrightarrow{a}||^2 + 4g(\overrightarrow{a}, \overrightarrow{n_1})^2}{4|n|||\overrightarrow{a}||g(\overrightarrow{a}, \overrightarrow{n_1})} & \textit{when } n(a_1 - a_2) > 0 \\ -\dfrac{|n|^2||\overrightarrow{a}||^2 + 4g(\overrightarrow{a}, \overrightarrow{n_1})^2}{4|n|||\overrightarrow{a}||g(\overrightarrow{a}, \overrightarrow{n_1})} & \textit{when } n(a_1 - a_2) < 0, \end{cases} \tag{35}$$

$$\sinh\theta(\overrightarrow{a}, \overrightarrow{n_1}) = \begin{cases} \dfrac{|n|^2||\overrightarrow{a}||^2 - 4g(\overrightarrow{a}, \overrightarrow{n_1})^2}{4|n|||\overrightarrow{a}||g(\overrightarrow{a}, \overrightarrow{n_1})} & \textit{when } n(a_1 - a_2) < 0 \\ -\dfrac{|n|^2||\overrightarrow{a}||^2 - 4g(\overrightarrow{a}, \overrightarrow{n_1})^2}{4|n|||\overrightarrow{a}||g(\overrightarrow{a}, \overrightarrow{n_1})} & \textit{when } n(a_1 - a_2) > 0; \end{cases} \tag{36}$$

$$\cosh\theta(\overrightarrow{n_2}, \overrightarrow{a}) = \begin{cases} \dfrac{|n|^2||\overrightarrow{a}||^2 + 4g(\overrightarrow{a}, \overrightarrow{n_2})^2}{4|n|||\overrightarrow{a}||g(\overrightarrow{a}, \overrightarrow{n_2})} & \textit{when } n(a_1 + a_2) > 0 \\ -\dfrac{|n|^2||\overrightarrow{a}||^2 + 4g(\overrightarrow{a}, \overrightarrow{n_2})^2}{4|n|||\overrightarrow{a}||g(\overrightarrow{a}, \overrightarrow{n_2})} & \textit{when } n(a_1 + a_2) < 0, \end{cases} \tag{37}$$

$$\sinh\theta(\overrightarrow{n_2}, \overrightarrow{a}) = \begin{cases} \dfrac{|n|^2||\overrightarrow{a}||^2 - 4g(\overrightarrow{a}, \overrightarrow{n_2})^2}{4|n|||\overrightarrow{a}||g(\overrightarrow{a}, \overrightarrow{n_2})} & \textit{when } n(a_1 + a_2) < 0 \\ -\dfrac{|n|^2||\overrightarrow{a}||^2 - 4g(\overrightarrow{a}, \overrightarrow{n_2})^2}{4|n|||\overrightarrow{a}||g(\overrightarrow{a}, \overrightarrow{n_2})} & \textit{when } n(a_1 + a_2) > 0. \end{cases} \tag{38}$$

Based on the definitions given above for the oriented Minkowskian pseudo-angles between any null vector and any spacelike or timelike vector, and also in view of (5), *the oriented Minkowskian pseudo-angle* $\theta(D_i, L)$ *between a null direction and any spacelike or timelike direction* or *between one of the null lines* D_1, D_2 *and any non-null line L passing through the origin of a Minkowskian plane* E_1^2, may be well defined as the oriented Minkowskian pseudo-angle $\theta(\overrightarrow{d}_i, \overrightarrow{l})$, whereby $\overrightarrow{l}$ is a unit vector on the line L; $\theta(D_i, L) = \theta(\overrightarrow{d}_i, \overrightarrow{l})$.

5. The Minkowskian Angles between Null Directions

Finally, to deal with the situation involving two null vectors, of course, two cases are to be considered: (i) *the null vectors are co-linear* and (ii) *they are not*. To begin with, *for the normalised null vectors* $\pm\overrightarrow{d}_i$, *the oriented Minkowskian angles* $\theta(\pm\overrightarrow{d}_i, \pm\overrightarrow{d}_j)$, $(i, j \in \{1,2\})$, *are defined to be given by their pseudo-angles* $\psi(\pm\overrightarrow{d}_i, \pm\overrightarrow{d}_j)$ *of Helzer*. In addition, based on (5) and (6), it then suffices to look at the angles $\theta(\overrightarrow{d}_1, \overrightarrow{d}_1)$, $\theta(\overrightarrow{d}_2, \overrightarrow{d}_2)$ and $\theta(\overrightarrow{d}_2, \overrightarrow{d}_1)$. (i) In the case of *normalised null vectors on the same diagonal*, since $\psi(\overrightarrow{v}, \overrightarrow{v}) = 0$ for all $\overrightarrow{v} \neq \overrightarrow{o}$,

$$\theta(\overrightarrow{d}_1, \overrightarrow{d}_1) = \theta(\overrightarrow{d}_2, \overrightarrow{d}_2) = 0. \tag{39}$$

(ii) In the case of *normalised null vectors on different diagonals*,

$$\theta(\overrightarrow{d}_2, \overrightarrow{d}_1) = \ln 2 + \ln 2 = 2\ln 2. \quad (40)$$

In case (i), matters are well in agreement with what we commonly expect to be natural enough the way it is, while, later on, there will follow some geometrical comments that may make the value $2\ln 2$ occurring in case (ii) not to appear as too unnatural after all. For the time being and for the sake of more easy reference, (39) and (40) will be put together in the following.

Theorem 5. *For the normalised null vectors $\overrightarrow{d_1} = (1,1)$ and $\overrightarrow{d_2} = (-1,1)$ their oriented Minkowskian angles are given by $\theta(\overrightarrow{d_1}, \overrightarrow{d_1}) = \theta(\overrightarrow{d_2}, \overrightarrow{d_2}) = 0$ and $\theta(\overrightarrow{d_2}, \overrightarrow{d_1}) = 2\ln 2$.*

At this stage, without further expanding on it, since (29) and (30) in particular imply that

$$\theta(\pm\overrightarrow{d_i}, \pm\overrightarrow{e_i}) = \pm\ln 2, \quad (41)$$

we may conclude the following.

Proposition 1. *In a Minkowskian plane E_1^2 the standard basic vectors $\{\pm\overrightarrow{e_1}, \pm\overrightarrow{e_2}\}$ are the only unit vectors which bisect the null vectors $\{\pm\overrightarrow{d_1}, \pm\overrightarrow{d_2}\}$.*

Next, for an arbitrary pair of null vectors, their oriented Minkowskian angle is defined as to be given by the oriented Minkowskian angle between their normalised null vectors. In addition, for any pair of null directions or diagonals D_i and D_j, their Minkowskian angle is defined to be the Minkowskian angle between their normalised vectors $\overrightarrow{d}_i$ and $\overrightarrow{d}_j$; $\theta(D_i, D_j) = \theta(\overrightarrow{d_i}, \overrightarrow{d_j})$.

6. The Unoriented Minkowskian Angles and Pseudo-Angles

For two *vectors $\overrightarrow{v}$ and $\overrightarrow{w}$* of any causal characters each and in whatever combination together, let $\theta(\overrightarrow{v}, \overrightarrow{w})$ be their *oriented Minkowskian angle* when both $\overrightarrow{v}$ and $\overrightarrow{w}$ are non-null vectors or when both are null vectors (cfr. Sections 3 and 5) or their *oriented Minkowskian pseudo-angle* when one of the vectors $\overrightarrow{v}$ and $\overrightarrow{w}$ is null and the other one is non-null (cfr. Section 4). In any case, from (6), it follows that $\theta(\overrightarrow{w}, \overrightarrow{v}) = -\theta(\overrightarrow{v}, \overrightarrow{w})$, so that it makes sense to define $\overline{\theta}(\overrightarrow{v}, \overrightarrow{w}) = |\theta(\overrightarrow{v}, \overrightarrow{w})| = |\theta(\overrightarrow{w}, \overrightarrow{v})|$ as the *unoriented or absolute Minkowskian angle or Minkowskian pseudo-angle between these vectors*. And, *the unoriented or absolute Minkowskian angles between two non-null directions and between two null directions* and *the unoriented or absolute Minkowskian pseudo-angles between one null and one non-null direction* are likewise defined.

In a Minkowskian plane E_1^2, geometrically, the two most distinguished directions may very well be *the null directions D_1 and D_2*; their *absolute Minkowskian angle* is given by $\overline{\theta}(D_1, D_2) = 2\ln 2$. The *absolute Minkowskian pseudo-angles between the co-ordinate axes A_1 and A_2* (spanned respectively by the standard unit vectors $\pm\overrightarrow{e_1}$ and $\pm\overrightarrow{e_2}$) *and the null diagonals D_1 and D_2* (spanned by $\pm\overrightarrow{d_1}$ and $\pm\overrightarrow{d_2}$) being given by $\overline{\theta}(A_i, D_j) = \ln 2$, $(i, j \in \{1,2\})$, and, further also taking into account (29) and (30), in a way, Proposition 1 may be reformulated as follows.

Proposition 2. *In a Minkowskian plane E_1^2, the co-ordinate axes A_1 and A_2 are geometrically characterised as the only two bisectrices of the null lines D_1 and D_2.*

7. A Geometrical Meaning of the Minkowskian Angles and Pseudo-Angles

Let us recall that *the Minkowskian arclengths L on the unit Minkowskian circle $H : z_1^2 - z_2^2 = \pm 1$*, say, for simplicity, *from $e_1 = (1,0)$ to the points $\overrightarrow{p} = (p_1, p_2)$ on its upper branch $H_2^+ : z_2 = (z_1^2 - 1)^{\frac{1}{2}}, z_1 \geq 1$*, are given by

$$
\begin{aligned}
L(\vec{e}_1, \vec{p}) &= \int_1^{p_1} |1-(dz_2/dz_1)^2|^{-1/2} dz_1 \\
&= \int_1^{p_1} (z_1^2-1)^{-1/2} dz_1 \\
&= \ln\{z_1+(z_1^2-1)^{1/2}\}\big|_1^{p_1} \\
&= \ln(p_1+p_2),
\end{aligned} \tag{42}
$$

cfr. Figure 7; (the readers likely will have thought about this already when observing the former formulae (2) and (3), from Helzer). And for geometrical interpretations of *oriented Minkowskian angles and pseudo-angles, it is good "to count" Minkowskian lengths on and of arcs on the Euclidean hyperbola's H taking into account their orientations as indicated in Figure* 8.

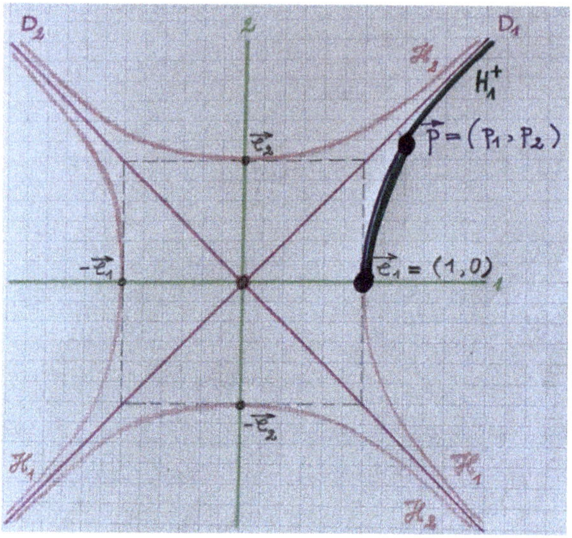

Figure 7. Arclengths on the unit circle.

Next, *Minkowskian pseudo-distances* $\tilde{d}$ *between two points on the same half straight null line* $z_2 = \pm z_1, z_1 > 0$ *or* $z_1 < 0$*, and the Minkowskian pseudo-lengths of the corresponding parts on the diagonals* D_1 *and* D_2 *are defined by the Minkowskian lengths of arcs on H that are determined thereupon by the Euclidean orthogonal projections of these points on* D_1 *or* D_2. To be more concrete, say, for points $\vec{q} = (q, q)$ and $\vec{r} = (r, r), (q, r \in R_0^+)$, on $D_1 : z_2 = z_1, z_1 > 0$, their Minkowskian pseudo-distance $\tilde{d}(\vec{q}, \vec{r})$ is defined by the Minkowskian arclength $L(\tilde{\vec{q}}, \tilde{\vec{r}})$ on H_1^+ between the Euclidean projections $\tilde{\vec{q}}$ and $\tilde{\vec{r}}$ on H_1^+ of the points $\vec{q}$ and $\vec{r}$ of D_1 orthogonal in the Euclidean sense to D_1, cfr. Figure 9:

$$
\begin{aligned}
\tilde{d}(\vec{q}, \vec{r}) &= L(\tilde{\vec{q}}, \tilde{\vec{r}}) \\
&= \bar{\theta}(\tilde{\vec{q}}, \tilde{\vec{r}}).
\end{aligned} \tag{43}
$$

And, finally, let us—in maybe too primitive a way—think $C = H \cup \{\pm \vec{d}_1, \pm \vec{d}_2\}$ as *a closed central curve*, centered at the origin O of the Minkowskian plane E_1^2, having precisely one point in each radial direction going out of O. Then, any pair of directions in this plane well determines a pair of points on C. In addition, *the oriented Minkowskian angles or pseudo-angles between these directions then correspond to the oriented arclengths on H and the oriented pseudo-lengths of parts of the null lines* D_1 *and* D_2*, whereby these pseudo-lengths come about in an oriented way as suggested in Figure* 10. By way of examples, here are the Minkowskian angles or pseudo-angles θ between some unit spacelike or timelike vectors and some normalised null vectors as well, whereby $\overrightarrow{u}_1 = (\frac{5}{4}, \frac{3}{4})$, $\overrightarrow{u}_2 = (\frac{3}{4}, \frac{5}{4})$, $\overrightarrow{u}_3 = (\frac{-3}{4}, \frac{5}{4})$ and $\overrightarrow{u}_4 = (\frac{-5}{4}, \frac{3}{4})$: $\theta(\overrightarrow{e}_1, \overrightarrow{d}_1) = \ln 2$, $\theta(\overrightarrow{u}_1, \overrightarrow{d}_1) = 0$, $\theta(\overrightarrow{d}_1, \overrightarrow{d}_2) = -2\ln 2$, $\theta(\overrightarrow{e}_1, \overrightarrow{u}_2) = \ln 2$, $\theta(\overrightarrow{e}_1, \overrightarrow{e}_2) = 0$, $\theta(\overrightarrow{e}_1, \overrightarrow{u}_3) = -\ln 2$, $\theta(\overrightarrow{d}_1, \overrightarrow{u}_2) = 0$, $\theta(\overrightarrow{d}_1, \overrightarrow{u}_3) = -2\ln 2$, $\theta(\overrightarrow{e}_1, \overrightarrow{u}_4) = -\ln 2$, $\theta(\overrightarrow{e}_1, \overrightarrow{p}) = \ln(p_1 + p_2)$, $\theta(\overrightarrow{e}_1, \overrightarrow{u}_1) = \ln 2$, $\theta(\overrightarrow{u}_1, \overrightarrow{p}_1) = \ln(p_1 + p_2) - \ln 2$; (cfr. Figure 11).

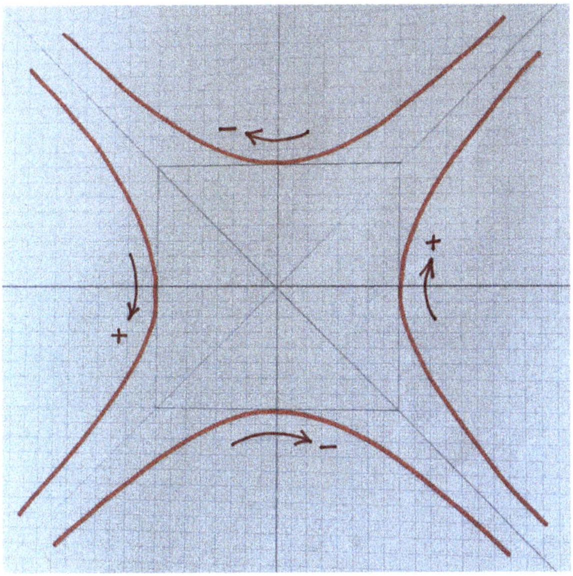

Figure 8. On orienting the unit circle.

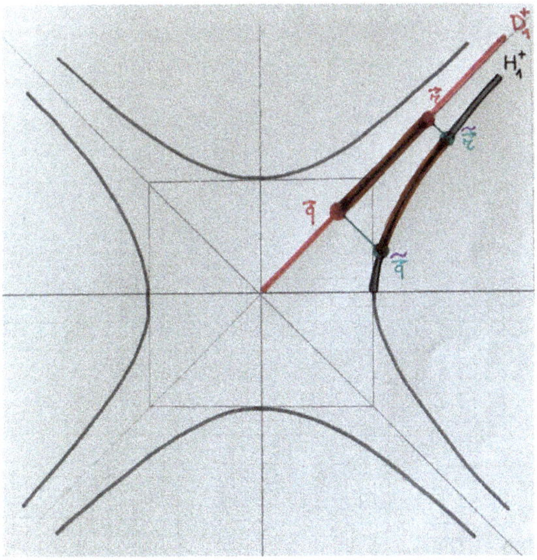

Figure 9. Minkowskian pseudo-lengths.

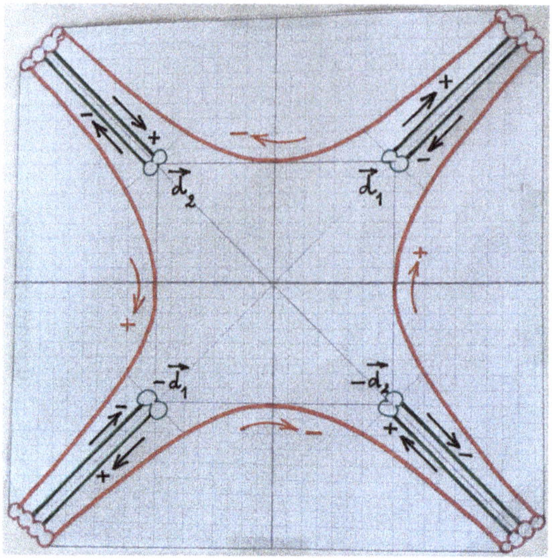

Figure 10. The central angles basic curve.

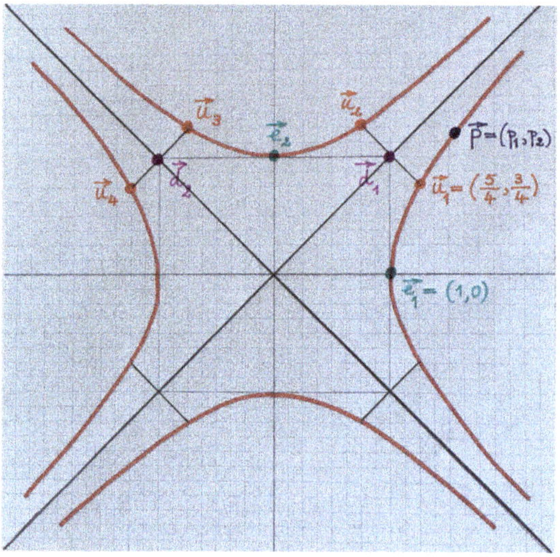

Figure 11. Some examples.

8. Conclusions

The 4D physical space-time of Minkowski with co-ordinates $(x, y, z; t)$ is the 4D pseudo-Euclidean geometrical space that is the product of a negative definite Euclidean line $(R, -dt^2)$ and a positive definite 3D Euclidean space $(R^3, dx^2 + dy^2 + dz^2)$, whereby the time-space scaling "i seconds = 300 000 kilometers" is taken into account. At any given moment of time t, the angles between any two directions in the physical 3D Euclidean (x, y, z) space at that moment are their standard original Euclidean angles; they are algebraically determined in terms of the group of the Euclidean rotations in a plane around a same point in this plane and they are geometrically measured by the Euclidean lengths of corresponding arcs on a Euclidean unit circle. From a natural scientific point of view there has been no immediate need to be occupied with looking for meaningful angles between two directions with arbitrary causal characters in planes of Minkowski. However, for two spacelike directions and geometrically equivalently for two timelike directions in a Minkowskian plane that belong to a same branch of the Minkowskian unit circle -an Euclidean orthogonal hyperbola-, their Minkowskian angles classically have been determined algebraically and measured geometrically by the straightforward adaptation of the traditional Euclidean approaches, now making use of the Minkowskian rotations with a same center and with Minkowskian lengths of corresponding arcs on a Minkowskian unit circle.

On the other hand, the algebraical definition in Minkowskian geometry of a proper notion of angle in the Euclidean way fails for directions that from a center point toward different branches of a Minkowskian unit circle with this center and also fails when null directions are involved. In the present paper, a geometrical generalisation of the Euclidean measure of angles between any two directions as the Euclidean lengths of corresponding arcs on a Euclidean unit circle is given for any two directions with arbitrary causal characters in a Minkowskian plane, by a well-defined notion of the Minkowskian angles or pseudo-angles of these two directions. This notion bases on the measurements of Minkowskian lengths and pseudo-lengths of corresponding parts of a Minkowskian unit circle and of parts of the asymptotes of this Euclidean orthogonal hyperbola. However imperfect that this extension of Euclidean angles to Minkowskian angles and pseudo-angles cannot help to be, it does have qualities of generality and of geometrical naturalness (up to an eventual change of calibration related to the choice of normalisation of the null vectors). And, of course, the classical Minkowskian angles between any two spacelike directions and between any two timelike directions within a same branch of the Minkowskian unit circle do properly fit in well into the above given notion of central Minkowskian angles and pseudo-angles.

Acknowledgments: The present paper originated in *discussions* with Miroslava Petrović-Torgašev and Emilija Nešović, at Kragujevac in 2014, and in our subsequent written *correspondence, about the approach of Helzer to pseudo-angles or "angles" between two vectors in a Minkowskian plane* [27]. Some ten years before, these Kragujevac geometers together with the author had defined and studied *the Minkowskian angles between any two non-null directions with different causal characters* [31]. *Our geometrical intuitive basis for this definition already then had been that of Minowskian central angles*. From [27,31], the contents of the present paper then came up, with, in particular, its geometrical interpretation of the herein newly defined Minkowskian pseudo-angles between any null direction and any non-null direction (spacelike and timelike alike) as *central Minkowskian pseudo-angles in terms of Minkowskian lengths and of Minkowskian pseudo-lengths*. In the meantime, several other colleagues also had been so kind to let me know their comments on this work and/or to suggest to include some related articles in the list of references of this paper, which indeed I did. In addition, I want to express my sincere thanks to all these geometers because I did learn something from all of them! Finally, for what some readers might very likely find a far too old fashioned or even untolerable geometrical look at the planes of Minkowski, I take full personal responsibility.

Radu Rosca did make significant contributions to several important fields in geometry. In particular, somewhat connected to the contents of the present paper, special mention could be made here of his *theories of pseudo-null or pseudo-isotropic submanifolds* (i.e., of non-degenerate submanifolds with ametrical Gauss maps) *in pseudo-Riemannian spaces* and *of nowadays so-called trapped submanifolds that have a non-trivial null mean curvature vector field*. I vividly remember the way that he introduced me, in particular, to his research in these beautiful chapters of the book of geometry, just as it were yesterday. However, only after having written the Introduction of this paper (whereby the year of "Raum und Zeit" made me think that it could be a good occasion for dedicating it to my first teacher of research in geometry), I did realise that this happened already almost half a century ago.

Conflicts of Interest: The author declares no conflict of interests.

References

1. Birman, G.S.; Desideri, G.M. *Una Introduction A La Geometria De Lorentz*; Conicet/Universidad Nacional Del Sur: Bahia Blanca, Argentina, 2012.
2. Chen, B.Y. *Pseudo-Riemaninan Geometry, δ-Invariants and Applications*; World Scientific: Hackensack, NJ, USA, 2011.
3. Chen, B.Y. *Differential Geometry of Warped Product Manifolds and Submanifolds*; World Scientific: Hackensack, NJ, USA, 2017.

4. Lopez, R. Differential geometry of curves and surfaces in Lorentz-Minkowski space. *Int. Electron. J. Geom.* **2014**, *7*, 44–107.
5. O'Neill, B. *Semi-Riemannian Geometry with Applications to Relativity*; Academic Press: New York, NY, USA, 1983.
6. Palomo, F.J.; Romero, A. *Certain actual topics on modern Lorentzian geometry*. In *Handbook of Differential Geometry Volume II*; Elsevier: Amsterdam, The Netherlands, 2006; pp. 513–546.
7. Romero, A. An Introduction to Certain Topics on Lorentzian Geometry. In *Topics in Modern Differential Geometry*; Atlantis Press: Paris, France; Springer: Berlin, Germany, 2017; pp. 259–284.
8. Vranceanu, G.; Rosca, R. *Introduction in Relativity and Pseudo-Riemannian Geometry*; Editura Academiei Republicii Socialiste Romania: Bucharest, Romania, 1976.
9. Yaglom, I.M. *A Simple Non-Euclidean Geometry In addition, Its Physical Basis*; Springer–Verlag: New York, NY, USA, 1979.
10. Ali, A.; Lopez, R. Slant helices in Minkowski space E_1^3. *J. Korean Math. Soc.* **2011**, *48*, 159–167.
11. Barros, M.; Caballero, M.; Ortega, M. Rotational surfaces in L^3 and solitons in the nonlinear σ–model. *Commun. Math. Phys.* **2009**, *290*, 437–477.
12. Barros, M.; Ferrandez, A. Null scrolls as fluctuating surfaces: A new simple way to cunstruct extrinsic string solutions. *J. High Energy Phys.* **2012**, *2012*, 68.
13. Barros, M.; Ferrandez, A. A new classical string solutions in AdS_3 through null scrolls. *Class. Quantum Grav.* **2013**, *30*, 115003.
14. Dillen, F.; Fastenakels, J.; Van der Veken, J.; Vrancken, L. Constant angle surfaces in $S^2 \times R$. *Monatsh. Math.* **2007**, *152*, 89–96.
15. Di Scala, A.J.; Ruiz-Hernandez, G. Helix submanifolds of Euclidean spaces. *Monatsh. Math.* **2009**, *157*, 205–215.
16. Ferrandez, A.; Gimenez, A.; Lucas, P. Null generalized helices in Lorentz-Minkowski spaces. *J. Phys. A Math. Gen.* **2002**, *35*, 8243–8251.
17. Haesen, S.; Nistor, A.I.; Verstraelen, L. On Growth and Form and Geometry I. *Kragujev. J. Math.* **2012**, *36*, 5–25.
18. Helzer, G. Relativity with acceleration. *Am. Math. Mon.* **2000**, *107*, 219–237.
19. Karadag, H.B.; Karadag, M. Null generalized slant helices in Lorentzian space. *Differ. Geom. Dyn. Syst.* **2008**, *10*, 178–185.
20. Lopez, R.; Munteanu, M.I. Constant angle surfaces in Minkowski space. *Bull. Belg. Math. Soc. Simon Stevin* **2011**, *18*, 271–286.
21. Munteanu, M.I. From golden spirals to constant slope surfaces. *J. Math. Phys.* **2010**, *51*, 073507.
22. Palmer, B. Bäcklund transformations for surfaces in Minkowski space. *J. Math. Phys.* **1990**, *31*, 2872.
23. Polyakov, A.M. Fine structure of strings. *Nucl. Phys. B* **1986**, *268*, 406–412.
24. Sahin, B.; Kilic, E.; Günes, R. Null helices in R_1^3. *Differ. Geom. Dyn. Syst.* **2001**, *3*, 31–36.
25. Şenol, A.; Ziplar, E.; Yayli, Y. Darboux helices in Minkowski space R_1^3. *Life Sci. J.* **2012**, *9*, 5905–5910.
26. Tian, C. Bäcklund transformation on surfaces with $K = -1$ in $R^{2,1}$. *J. Geom. Phys.* **1997**, *22*, 212–218.
27. Helzer, G. A relativistic version of the Gauss–Bonnet formula. *J. Differ. Geom.* **1974**, *9*, 507–512.
28. Birman, G.; Nomizu, K. Trigonometry in Lorentzian geometry. *Am. Math. Mon.* **1984**, *91*, 543–549.
29. Birman, G.; Nomizu, K. The Gauss-Bonnet theorem for 2-dimensional space-times. *Mich. Math. J.* **1984**, *31*, 77–81.
30. Nešović, E.; Petrović-Torgašev, M. Some trigonometric relations in the Lorentzian plane. *Kragujev. J. Math.* **2003**, *25*, 219–225.
31. Nešović, E.; Petrović-Torgašev, M.; Verstraelen, L. Curves in Lorentzian spaces. *Boll. Unione Mat. Ital.* **2005**, *8*, 685–696.

mathematics

Article

Generic Properties of Framed Rectifying Curves

Yongqiao Wang [1], Donghe Pei [2,*] and Ruimei Gao [3]

1 School of Science, Dalian Maritime University, Dalian 116026, China; wangyq794@nenu.edu.cn
2 School of Mathematics and Statistics, Northeast Normal University, Changchun 130024, China
3 Department of Science, Changchun University of Science and Technology, Changchun 130022, China; gaorm135@nenu.edu.cn
* Correspondence: peidh340@nenu.edu.cn; Tel.: +86-431-8509-9155

Received: 27 November 2018; Accepted: 26 December 2018; Published: 3 January 2019

Abstract: The position vectors of regular rectifying curves always lie in their rectifying planes. These curves were well investigated by B.Y.Chen. In this paper, the concept of framed rectifying curves is introduced, which may have singular points. We investigate the properties of framed rectifying curves and give a method for constructing framed rectifying curves. In addition, we reveal the relationships between framed rectifying curves and some special curves.

Keywords: framed rectifying curves; singular points; framed helices; centrodes; circular rectifying curves

MSC: 53A04; 57R45; 58K05

1. Introduction

Curves, which are the basic objects of study, have attracted much attention from many mathematicians and physicists [1–3]. Due to the need to observe the properties of special curves, a renewed interest in curves has developed, such as rectifying curves in different spaces. The space curves whose position vectors always lie in their rectifying planes are called rectifying curves. B.Y. Chen gave the notion of rectifying curves in [4]. In [5], the relationship between centrodes of space curves and rectifying curves was revealed by F. Dillen and B.Y. Chen. In kinematics, the centrode is the path traced by the instantaneous center of rotation of a rigid plane figure moving in a plane, and it has wide applications in mechanics and joint kinematics (see [6–9]).

Since B.Y. Chen's important work, the notion of rectifying curves was extended to other ambient spaces [10–13]. As we know, regular curves determine the curvature functions and torsion functions, which can provide valuable geometric information about the curves by the Frenet frames of the original curves. If space curves have singular points, the Frenet frames of these curves cannot be constructed. However, S. Honda and M. Takahashi [14] gave the definition of framed curves. Framed curves are space curves with moving frames, and they may have singular points. They are the generalizations of not only Legendrian curves in unit tangent bundles, but also regular curves with linear independent conditions (see [15]).

Inspired by the above work, in order to investigate the properties of rectifying curves with singular points, we should give the concept of framed rectifying curves. The difficulties arise because tangent vectors vanish at singular points, so it is impossible to normalize tangent vectors, principal normal vectors, and binormal vectors in the usual way. Here, we define the generalized tangent vector, the generalized principle normal vector, and the generalized binormal vector, respectively. Actually, at regular points, they are just the usual tangent vector, principle vector, and binormal vector. We obtain moving adapted frames for framed rectifying curves, and some smooth functions similar to the curvature of regular curves are defined by using moving adapted frames. These functions are referred to as framed curvature, which is very useful to analyze framed rectifying curves. On this basis, we investigate the properties of framed rectifying curves and give some sufficient and necessary conditions for the judgment of framed rectifying curves. Moreover, we give a method for constructing framed rectifying curves. In this paper, framed helices are also defined. We discuss the relationship between framed rectifying curves and framed helices in terms of the ratio of framed curvature. In particular, the ratio of framed curvature for framed rectifying curves has extrema at singular points. In addition, we give the notions of the centrodes of the framed curves and circular rectifying curves and reveal the relationships between framed rectifying curves and these special curves.

The organization of this paper is as follows. We review the concept of the framed curve and define an adapted frame and framed curvature for the framed curve in Section 2. We provide some sufficient and necessary conditions for the judgment of framed rectifying curves in Section 3. An important result, which explicitly determines all framed rectifying curves, is given in Section 4. Moreover, the relationships between framed rectifying curves and framed helices and framed rectifying curves and centrodes are given in Sections 5 and 6, respectively. At last, we consider the contact between framed rectifying curves and model curves (circular rectifying curves) in Section 7.

2. Framed Curve and Adapted Frame

Let $\mathbb{R}^3$ be the three-dimensional Euclidean space, and let $\gamma : I \to \mathbb{R}^3$ be a curve with singular points. In order to investigate this curve, we will introduce the framed curve (cf., [14]). We denote the set Δ_2 as follows:

$$\Delta_2 = \{\boldsymbol{\mu} = (\boldsymbol{\mu}_1, \boldsymbol{\mu}_2) \in \mathbb{R}^3 \times \mathbb{R}^3 | \boldsymbol{\mu}_i \cdot \boldsymbol{\mu}_j = \delta_{ij},\ i, j = 1, 2\}.$$

Then, Δ_2 is a three-dimensional smooth manifold. Let $\boldsymbol{\mu} = (\boldsymbol{\mu}_1, \boldsymbol{\mu}_2) \in \Delta_2$. We define a unit vector $\boldsymbol{\nu} = \boldsymbol{\mu}_1 \times \boldsymbol{\mu}_2$ in $\mathbb{R}^3$. This means that $\boldsymbol{\nu}$ is orthogonal to $\boldsymbol{\mu}_1$ and $\boldsymbol{\mu}_2$.

Definition 1. *We say that $(\gamma, \boldsymbol{\mu}) : I \to \mathbb{R}^3 \times \Delta_2$ is a framed curve if $\langle \gamma'(s), \boldsymbol{\mu}_i(s) \rangle = 0$ for all $s \in I$ and $i = 1, 2$. We also say that $\gamma : I \to \mathbb{R}^3$ is a framed base curve if there exists $\boldsymbol{\mu} : I \to \Delta_2$ such that $(\gamma, \boldsymbol{\mu})$ is a framed curve.*

Let $(\gamma, \boldsymbol{\mu}_1, \boldsymbol{\mu}_2) : I \to \mathbb{R}^3 \times \Delta_2$ be a framed curve and $\boldsymbol{\nu}(s) = \boldsymbol{\mu}_1(s) \times \boldsymbol{\mu}_2(s)$. Then, we have the following Frenet–Serret formula:

$$\begin{cases} \boldsymbol{\mu}_1'(s) = l(s)\boldsymbol{\mu}_2(s) + m(s)\boldsymbol{\nu}(s) \\ \boldsymbol{\mu}_2'(s) = -l(s)\boldsymbol{\mu}_1(s) + n(s)\boldsymbol{\nu}(s) \\ \boldsymbol{\nu}'(s) = -m(s)\boldsymbol{\mu}_1(s) - n(s)\boldsymbol{\mu}_2(s). \end{cases}$$

Here, $l(s) = \langle \boldsymbol{\mu}_1'(s), \boldsymbol{\mu}_2(s) \rangle$, $m(s) = \langle \boldsymbol{\mu}_1'(s), \boldsymbol{\nu}(s) \rangle$ and $n(s) = \langle \boldsymbol{\mu}_2'(s), \boldsymbol{\nu}(s) \rangle$. In addition, there exists a smooth mapping $\alpha : I \to \mathbb{R}$ such that:

$$\gamma'(s) = \alpha(s)\boldsymbol{\nu}(s).$$

The four functions $(l(s), m(s), n(s), \alpha(s))$ are called the curvature of γ. If $m(s) = n(s) = 0$, then $\boldsymbol{\nu}'(s) = \mathbf{0}$. In this paper, we consider the case $\boldsymbol{\nu}'(s) \neq \mathbf{0}$. Obviously, $\alpha(s_0) = 0$ if and only if s_0 is a singular point of γ. We can use the curvature of the framed curve to analyze the singular points.

In [14], the theorems of the existence and uniqueness for framed curves were shown as follows:

Theorem 1. *Let $(l, m, n, \alpha) : I \to \mathbb{R}^4$ be a smooth mapping. There exists a framed curve $(\gamma, \boldsymbol{\mu}) : I \to \mathbb{R}^3 \times \Delta_2$ whose associated curvature of the framed curve is (l, m, n, α).*

Theorem 2. *Let $(\gamma, \boldsymbol{\mu})$ and $(\overline{\gamma}, \overline{\boldsymbol{\mu}}) : I \to \mathbb{R}^3 \times \Delta_2$ be framed curves whose curvatures of the framed curves (l, m, n, α) and $(\overline{l}, \overline{m}, \overline{n}, \overline{\alpha})$ coincide. Then, $(\gamma, \boldsymbol{\mu})$ and $(\overline{\gamma}, \overline{\boldsymbol{\mu}})$ are congruent as framed curves.*

Let $(\gamma, \mu_1, \mu_2) : I \to \mathbb{R}^3 \times \Delta_2$ be a framed curve with the curvature $(l(s), m(s), n(s), \alpha(s))$. μ_1 and μ_2 are the base vectors of the normal plane of $\gamma(s)$, as a case similar to the Bishop frame for regular curves [16]. We define $(\overline{\mu}_1, \overline{\mu}_2) \in \Delta_2$ by:

$$\begin{pmatrix} \overline{\mu}_1(s) \\ \overline{\mu}_2(s) \end{pmatrix} = \begin{pmatrix} \cos\theta(s) & -\sin\theta(s) \\ \sin\theta(s) & \cos\theta(s) \end{pmatrix} \begin{pmatrix} \mu_1(s) \\ \mu_2(s) \end{pmatrix}.$$

Here, $\theta(s)$ is a smooth function. Obviously, $(\gamma, \overline{\mu}_1, \overline{\mu}_2) \to \mathbb{R}^3 \times \Delta_2$ is also a framed curve, and we have:

$$\overline{\nu}(s) = \mu_1(s) \times \mu_2(s) = \overline{\mu}_1(s) \times \overline{\mu}_2(s) = \nu(s).$$

By straightforward calculations, we have:

$$\begin{aligned} \overline{\mu}_1'(s) =& (l(s) - \theta'(s)) \sin\theta(s)\mu_1(s) + (l(s) - \theta'(s)) \cos\theta(s)\mu_2(s) \\ &+ (m(s)\cos\theta(s) - n(s)\sin\theta(s))\nu(s), \end{aligned}$$

$$\begin{aligned} \overline{\mu}_2'(s) =& -(l(s) - \theta'(s)) \cos\theta(s)\mu_1(s) + (l(s) - \theta'(s)) \sin\theta(s)\mu_2(s) \\ &+ (m(s)\sin\theta(s) + n(s)\cos\theta(s))\nu(s). \end{aligned}$$

Let $\theta : I \to \mathbb{R}$ be a smooth function that satisfies $m(s)\sin\theta(s) = -n(s)\cos\theta(s)$. Assume that $m(s) = -p(s)\cos\theta(s)$, $n(s) = p(s)\sin\theta(s)$, then we have:

$$\nu'(s) = -m(s)\mu_1(s) - n(t)\mu_2(s) = p(s)(\cos\theta(s)\mu_1(s) - \sin\theta(s)\mu_2(s)) = p(s)\overline{\mu}_1(s),$$

$$\begin{aligned} \overline{\mu}_1'(s) =& (l(s) - \theta'(s)) \sin\theta(s)\mu_1(s) + (l(s) - \theta'(s)) \cos\theta(s)\mu_2(s) + (m(s)\cos\theta(s) - n(s)\sin\theta(s))\nu(s) \\ =& -p(s)\nu(s) + (l(s) - \theta'(s))\overline{\mu}_2(s) \end{aligned}$$

and:

$$\begin{aligned} \overline{\mu}_2'(s) =& -(l(s) - \theta'(s)) \cos\theta(s)\mu_1(s) + (l(s) - \theta'(s)) \sin\theta(s)\mu_2(s) + (m(s)\sin\theta(s) + n(s)\cos\theta(s))\nu(s) \\ =& -(l(s) - \theta'(s))\overline{\mu}_1(s). \end{aligned}$$

The vectors $\nu(s), \overline{\mu}_1(s), \overline{\mu}_2(s)$ form an adapted frame along $\gamma(s)$, and we have the following Frenet–Serret formula:

$$\begin{pmatrix} \nu'(s) \\ \overline{\mu}_1'(s) \\ \overline{\mu}_2'(s) \end{pmatrix} = \begin{pmatrix} 0 & p(s) & 0 \\ -p(s) & 0 & q(s) \\ 0 & -q(s) & 0 \end{pmatrix} \begin{pmatrix} \nu(s) \\ \overline{\mu}_1(s) \\ \overline{\mu}_2(s) \end{pmatrix}.$$

We call the vectors $\nu(s), \overline{\mu}_1(s), \overline{\mu}_2(s)$ the generalized tangent vector, the generalized principle normal vector, and the generalized binormal vector of the framed curve, respectively, where $p(s) = |\nu'(s)| > 0$ and $q(s) = l(s) - \theta'(s)$. The functions $(p(s), q(s), \alpha(s))$ are referred to as the framed curvature of $\gamma(s)$.

Proposition 1. *Let $(\gamma, \overline{\mu}_1, \overline{\mu}_2) : I \to \mathbb{R}^3 \times \Delta_2$ be a framed curve. The relationships among the curvature $\kappa(s)$, the torsion $\tau(s)$, and the framed curvature $(p(s), q(s), \alpha(s))$ of a regular curve are given by:*

$$\kappa(s) = \frac{p(s)}{|\alpha(s)|}, \quad \tau(s) = \frac{q(s)}{\alpha(s)}.$$

Proof. By straightforward calculations, we have:

$$\gamma'(s) = \alpha(s)\nu(s),$$

$$\gamma''(s) = \alpha'(s)\boldsymbol{\nu}(s) + \alpha(s)p(s)\overline{\boldsymbol{\mu}}_1(s),$$

$$\gamma'''(s) = (\alpha''(s) - \alpha(s)p^2(s))\boldsymbol{\nu}(s) + (2\alpha'(s)p(s) + \alpha(t)p'(s))\overline{\boldsymbol{\mu}}_1(s) + \alpha(s)p(s)q(s)\overline{\boldsymbol{\mu}}_2(s).$$

It follows:

$$|\gamma'(s)| = |\alpha(s)|,$$

$$|\gamma'(s) \times \gamma''(s)| = \alpha^2(s)p(s),$$

$$\det(\gamma'(s), \gamma''(s), \gamma'''(s)) = \alpha^3(s)p^2(s)q(s).$$

Therefore, the relationships are shown by:

$$\kappa(s) = \frac{|\gamma'(s) \times \gamma''(s)|}{|\gamma'(s)|^3} = \frac{p(s)}{|\alpha(s)|},$$

$$\tau(s) = \frac{\det(\gamma'(s), \gamma''(s), \gamma'''(s))}{|\gamma'(s) \times \gamma''(s)|^2} = \frac{q(s)}{\alpha(s)}.$$

□

3. Framed Rectifying Curves

In this section, the framed rectifying curves are defined, and we investigate their properties.

Definition 2. *Let $(\gamma, \overline{\boldsymbol{\mu}}_1, \overline{\boldsymbol{\mu}}_2) : I \to \mathbb{R}^3 \times \Delta_2$ be a framed curve. We call γ a framed rectifying curve if its position vector γ satisfies:*

$$\gamma(s) = \lambda(s)\boldsymbol{\nu}(s) + \xi(s)\overline{\boldsymbol{\mu}}_2(s)$$

for some functions $\lambda(s)$ and $\xi(s)$.

Some properties of the framed rectifying curves are shown in the following theorem.

Theorem 3. *Let $(\gamma, \overline{\boldsymbol{\mu}}_1, \overline{\boldsymbol{\mu}}_2) : I \to \mathbb{R}^3 \times \Delta_2$ be a framed curve with $p(s) > 0$. The following statements are equivalent.*

(i) The relation between the framed curvature and the framed curve is as follows:

$$\langle \gamma(s), \boldsymbol{\nu}(s) \rangle' = \alpha(s).$$

(ii) The distance squared function satisfies $f(s) = \langle \gamma(s), \gamma(s) \rangle = \langle \gamma(s), \boldsymbol{\nu}(s) \rangle^2 + C$ for some positive constant C.

(iii) $\langle \gamma(s), \overline{\boldsymbol{\mu}}_2(s) \rangle = \xi$, ξ is a constant.

(iv) $\gamma(s)$ is a framed rectifying curve.

Proof. Let $\gamma(s)$ be a framed rectifying curve. By definition, there exist some functions $\lambda(s)$ and $\xi(s)$ such that:

$$\gamma(s) = \lambda(s)\boldsymbol{\nu}(s) + \xi(s)\overline{\boldsymbol{\mu}}_2(s). \tag{1}$$

By using the Frenet–Serret formula and taking the derivative of (1) with respect to s, we have:

$$\lambda'(s) = \alpha(s), \quad \lambda(s)p(s) = \xi(s)q(s), \quad \xi'(s) = 0. \tag{2}$$

From the first and third equalities of (2), we have that $\langle\gamma(s), \boldsymbol{\nu}(s)\rangle' = \lambda'(s) = \alpha(s)$. This proves Statement (i). Since $\xi'(s) = 0$, we can obtain Statement (iii). From (1) and (2), we have that $\langle\gamma(s), \gamma(s)\rangle = \lambda^2(s) + \xi^2 = \langle\gamma(s), \boldsymbol{\nu}(s)\rangle^2 + C$, $C = \xi^2$ is positive. This proves Statement (ii).

Conversely, let us assume that Statement (i) holds.

$$\langle\gamma(s), \boldsymbol{\nu}(s)\rangle' = \langle\alpha(s)\boldsymbol{\nu}(s), \boldsymbol{\nu}(s)\rangle + p(s)\langle\gamma(s), \overline{\boldsymbol{\mu}}_1(s)\rangle = \alpha(s).$$

Since $p(s) > 0$, by assumption, we have $\langle\gamma(s), \overline{\boldsymbol{\mu}}_1(s)\rangle = 0$. This means the curve is a framed rectifying curve.

If Statement (ii) holds, $\langle\gamma(s), \gamma(s)\rangle = \langle\gamma(s), \boldsymbol{\nu}(s)\rangle^2 + C$, where C is a positive constant. Then, we have:

$$2\langle\gamma(s), \alpha(s)\boldsymbol{\nu}(s)\rangle = 2\langle\gamma(s), \boldsymbol{\nu}(s)\rangle(\alpha(s) + p(s)\langle\gamma(s), \overline{\boldsymbol{\mu}}_1(s)\rangle)$$

and $\langle\gamma(s), \overline{\boldsymbol{\mu}}_1(s)\rangle = 0$. Therefore, $\gamma(s)$ is a framed rectifying curve. Statement (iii) implies that the curve is a framed rectifying curve by an appeal to the Frenet–Serret formula. □

Remark 1. *s_0 is a singular point of the framed rectifying curve γ if and only if $\alpha(s_0) = 0$. From (2) and Statement (ii), we know that the ratio $q(s)/p(s)$ and the distance squared function $f(s)$ have extrema at s_0.*

4. Construction Approach of Framed Rectifying Curves

In [4], the construction approach of regular rectifying curves is given by B. Y. Chen in Theorem 3, but it is not suitable for the non-regular case. In this section, a new construction approach is provided, which can be applied to both regular rectifying curves and non-regular rectifying curves. Moreover, it explicitly determines all framed rectifying curves in Euclidean three-space. First, we introduce the notion of the framed spherical curve.

Definition 3. *Let $(\gamma, \overline{\boldsymbol{\mu}}_1, \overline{\boldsymbol{\mu}}_2) : I \to \mathbb{R}^3 \times \Delta_2$ be a framed curve. We call γ a framed spherical curve if the framed base curve γ is a curve on S^2.*

We show the key theorem in this section as follows.

Theorem 4. *Let $(\gamma, \overline{\boldsymbol{\mu}}_1, \overline{\boldsymbol{\mu}}_2) : I \to \mathbb{R}^3 \times \Delta_2$ be a framed curve with $p(s) > 0$. Then, γ is a framed rectifying curve if and only if:*

$$\gamma(s) = \rho(\tan^2(\int |\boldsymbol{g}'(s)|ds + C) + 1)^{\frac{1}{2}}\boldsymbol{g}(s), \tag{3}$$

where C is a constant, ρ is a positive number, and $\boldsymbol{g}(s)$ is a framed spherical curve.

Proof. Let γ be a framed rectifying curve. From Theorem 3, we have $\langle\gamma(s), \gamma(s)\rangle = \lambda^2(s) + \rho^2$, where ρ is a positive number. The framed rectifying curve $\gamma(s)$ can be written as:

$$\gamma(s) = (\lambda^2(s) + \rho^2)^{\frac{1}{2}}\boldsymbol{g}(s), \tag{4}$$

where $g(s)$ is a framed spherical curve. By taking the derivative of (4), we have:

$$\gamma'(s) = \frac{\lambda(s)\alpha(s)}{(\lambda^2(s)+\rho^2)^{\frac{1}{2}}} g(s) + (\lambda^2(s)+\rho^2)^{\frac{1}{2}} g'(s). \tag{5}$$

As $\gamma'(s) = \alpha(s)\nu(s)$, $g'(s)$ is orthogonal to $g(s)$. Therefore, Equality (5) implies:

$$|g'(s)| = |\frac{\rho\alpha(s)}{\lambda^2(s)+\rho^2}|,$$

and we have

$$\int |\gamma'(s)| ds + C = \arctan(\frac{\lambda(s)}{\rho}).$$

Then, $\lambda(s) = \rho \tan(\int |g'(s)| ds + C)$, and substituting this equality into (4) yields (3).

Conversely, assume $\gamma(s)$ is a framed curve defined by:

$$\gamma(s) = \rho(\tan^2(\int |g'(s)| ds + C) + 1)^{\frac{1}{2}} g(s) \tag{6}$$

for a constant C, a positive number ρ, and a framed curve $g(s)$ on S^2. Let $\widetilde{\lambda}(s) = \rho \tan(\int |g'(s)| ds + C)$ and $\widetilde{\lambda}'(s) = \widetilde{\alpha}'(s)$. Then, $\int |g'(s)| ds + C = \arctan(\frac{\widetilde{\lambda(s)}}{\rho})$. By taking the derivative of this equality, we get:

$$\frac{\rho\widetilde{\alpha}(s)}{\widetilde{\lambda}^2(s)+\rho^2} = |g'(s)| \tag{7}$$

and:

$$\gamma'(s) = \frac{\widetilde{\lambda}(s)\widetilde{\alpha}(s)}{(\widetilde{\lambda}^2(s)+\rho^2)^{\frac{1}{2}}} g(s) + (\widetilde{\lambda}^2(s)+\rho^2)^{\frac{1}{2}} g'(s). \tag{8}$$

Equality (7) and Equality (8) imply that $|g'(s)| = \widetilde{\alpha}(s)$, since $g'(s) = \lambda(s)\nu(s)$. We have $\widetilde{\alpha}(s) = \pm\lambda(s), \widetilde{\lambda}(s) = \pm\lambda(s)$. Then:

$$\gamma(s) = (\lambda^2(s)+\rho^2)^{\frac{1}{2}} g(s), \tag{9}$$

which shows that the distance squared function satisfies Statement (ii) in Theorem 3. It follows that $\gamma(s)$ is a framed rectifying curve. □

Framed rectifying curves include regular rectifying curves and non-regular rectifying curves. We will give two examples.

Example 1. *Let* $g_1(s) = (\frac{1}{2}\cos 2s, \frac{1}{2}\sin 2s, \frac{\sqrt{3}}{2})$, $s \in (-\frac{\pi}{2}, \frac{\pi}{2})$, *then* $g_1(s)$ *is a space curve on* S^2. *We have* $|g_1'(s)| = 1$. *Let* $\rho = 1$ *and* $C = 0$. *By Theorem 4, we know that the curve:*

$$\gamma_1(s) = (\frac{\cos 2s}{2\cos s}, \sin s, \frac{\sqrt{3}}{2\cos s}), \quad s \in (-\frac{\pi}{2}, \frac{\pi}{2})$$

is a regular rectifying curve in $\mathbb{R}^3$ *(Figure 1).*

If $\gamma(s)$ *is a framed curve with singular points, this is different from the case that* $\gamma(s)$ *is a regular curve.*

Example 2. *Let* $g_2(s) = (\cos s^2 \cos s^3, \sin s^2 \cos s^3, \sin s^3)$, *then* $g_2(s)$ *is a curve in* S^2 *and* $|g_2'(s)| = (4s^2\cos^2 s^3 + 9s^4)^{\frac{1}{2}}$. *Let* $\rho = 1$ *and* $C = 0$. *By Theorem 4, we know that the curve:*

$$\gamma_2(s) = (\tan^2(\int (4s^2\cos^2 s^3 + 9s^4)^{\frac{1}{2}} ds) + 1)^{\frac{1}{2}} (\cos s^2 \cos s^3, \sin s^2 \cos s^3, \sin s^3)$$

is a framed rectifying curve with a cusp in $\mathbb{R}^3$ *(Figure 2).*

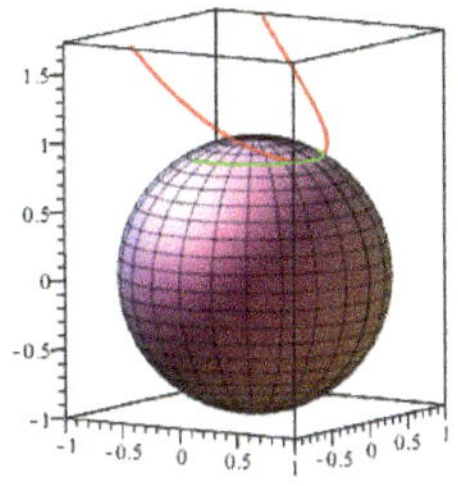

Figure 1. The red curve $\gamma_1(s)$ is the regular rectifying curve, and the green curve $g_1(s)$ is a curve on S^2.

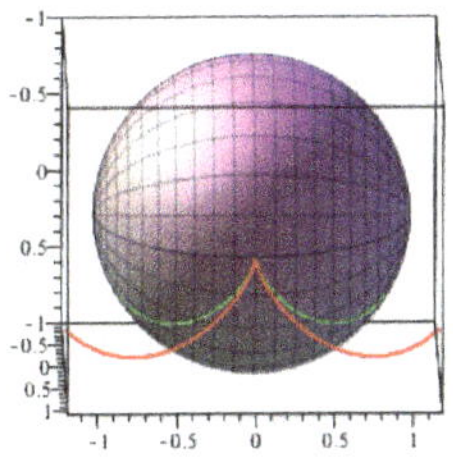

Figure 2. The red curve $\gamma_2(s)$ is the framed rectifying curve, and the green curve $g_2(s)$ is a curve on S^2.

5. Framed Rectifying Curves versus Framed Helices

In this section, we define the framed helices and investigate the relations between framed helices and framed rectifying curves.

Definition 4. *Let* $(\gamma, \overline{\mu}_1, \overline{\mu}_2) : I \to \mathbb{R}^3 \times \Delta_2$ *be a framed curve with* $p(s) > 0$. *We call* γ *a framed helix if there exists a fixed unit vector* ζ *satisfying:*

$$\langle \nu(s), \zeta \rangle = \cos\omega$$

for some constant ω.

We now consider the ratio $(q/p)(s)$ of the framed helix.

$$\langle \nu(s), \zeta \rangle = \cos\omega. \tag{10}$$

By taking the derivative of (10), as $p(s) > 0$ and $\langle \nu(s), \zeta \rangle' = p(s)\langle \overline{\mu}_1(s), \zeta \rangle$, we have:

$$\langle \overline{\mu}_1(s), \zeta \rangle = 0. \tag{11}$$

We know that ζ is in the plane whose basis vectors are $\nu(s)$ and $\overline{\mu}_2(s)$. As $\langle \nu(s), \zeta \rangle = \cos\omega$, we have $\langle \overline{\mu}_2(s), \zeta \rangle = \pm\sin\omega$. By taking the derivative of (11), we get:

$$-p(s)\langle \nu(s), \zeta \rangle + q(s)\langle \overline{\mu}_2(s), \zeta \rangle = 0,$$

then:

$$\frac{q(s)}{p(s)} = \pm\cot\omega. \tag{12}$$

For framed rectifying curves, a simple characterization in terms of the ratio $q(s)/p(s)$ is shown in the following theorem.

Theorem 5. *Let $(\gamma, \overline{\mu}_1, \overline{\mu}_2) : I \to \mathbb{R}^3 \times \Delta_2$ be a framed curve with $p(s) > 0$, then $\gamma(s)$ is a framed rectifying curve if and only if $q(s)/p(s) = c_1 \int \alpha(s)ds + c_2$ for some constants c_1 and c_2, with $c_1 \neq 0$.*

Proof. The proof is similar to that of Theorem 2 in [4]. If $\gamma(s)$ is a framed rectifying curve, from (2), we have that $q(s)/p(s) = \lambda(s)/\xi(s) = \lambda(s)/\xi$ for some constant ξ. Since $\lambda'(s) = \alpha(s)$ and $\xi \neq 0$, then the ratio of $q(s)$ and $p(s)$ satisfies $q(s)/p(s) = c_1 \int \alpha(s)ds + c_2$ for some constants c_1 and c_2, with $c_1 \neq 0$.

Conversely, suppose that $(\gamma, \overline{\mu}_1, \overline{\mu}_2) : I \to \mathbb{R}^3 \times \Delta_2$ is a framed curve with $p(s) > 0$, and $q(s)/p(s) = c_1 \int \alpha(s)ds + c_2$ for some constants c_1 and c_2, with $c_1 \neq 0$. If we put $\xi = 1/c_1$ and $\lambda(s) = \int \alpha(s)ds + c_2/c_1$, hence, by invoking the Frenet–Serret formula, we obtain:

$$\frac{d}{ds}[\gamma(s) - \lambda(s)\boldsymbol{\nu}(s) - \xi\overline{\boldsymbol{\mu}}_2(s)] = (\xi q(s) - \lambda(s)p(s))\overline{\boldsymbol{\mu}}_1(s) = 0.$$

This means that $\gamma(s)$ is congruent to a framed rectifying curve. □

Remark 2. *If γ is a framed rectifying curve, we have $\lambda(s)p(s) = \xi q(s)$ for some constant ξ. If $\xi = 0$, then $\lambda(s)p(s) = 0$, as $p(s) > 0$, so $\lambda(s) \equiv 0$. This means that $\gamma(s)$ is a point.*

After that, we reveal the relationship between the framed rectifying curves and the framed helices. We have the following theorem:

Theorem 6. *Let $(\gamma, \overline{\mu}_1, \overline{\mu}_2) : I \to \mathbb{R}^3 \times \Delta_2$ be a framed curve with $p(s) > 0$, the framed curvature functions satisfying $(q/p)(s) = c_1 \int \alpha(s)ds + c_2$, for some constants c_1 and c_2. If $c_1 = 0$, we will get framed helices; otherwise, we get framed rectifying curves.*

6. Framed Rectifying Curves versus Centrodes

The centrodes play important roles in joint kinematics and mechanics (see [5]). We can define the centrodes of framed curves. For a framed curve γ in $\mathbb{R}^3$, the curve defined by the vector $\boldsymbol{d} = q\boldsymbol{\nu} + p\overline{\boldsymbol{\mu}}_2$, which is called the centrode of framed curve γ.

The following results establish some relationships between framed rectifying curves and centrodes.

Theorem 7. *The centrode of a framed curve with nonzero constant framed curvature function $p(s)$ and nonconstant framed curvature function $q(s)$ is a framed rectifying curve. Conversely, the framed rectifying curve in $\mathbb{R}^3$ is the centrode of some framed curve with nonconstant framed curvature function $q(s)$ and nonzero constant framed curvature function $p(s)$.*

Proof. Let $\gamma(s)$ be a framed curve with nonzero constant framed curvature $p(s)$ and nonconstant framed curvature $q(s)$. Consider the centrode of $\gamma(s)$:

$$\boldsymbol{d}(s) = q(s)\boldsymbol{\nu}(s) + p(s)\overline{\boldsymbol{\mu}}_2(s).$$

$\boldsymbol{d}(s)$ can also be seen as a framed curve. Let the vectors $\overline{\boldsymbol{\mu}}_{d,1}(s), \overline{\boldsymbol{\mu}}_{d,2}(s), \boldsymbol{\nu}_d(s)$ be the adapted frame along $\boldsymbol{d}(s)$. By differentiating the centrode, then we have $\boldsymbol{d}'(s) = q'(s)\boldsymbol{\nu}(s)$, which implies that unit vector $\boldsymbol{\nu}_d(s)$ and unit vector $\boldsymbol{\nu}(s)$ at the corresponding points are parallel. Then, the first equality in Frenet–Serret formula implies that $\overline{\boldsymbol{\mu}}_{d,1}(s)$ and $\overline{\boldsymbol{\mu}}_1(s)$ at the corresponding points are also parallel. Hence, $\overline{\boldsymbol{\mu}}_{d,2}(s)$ and $\overline{\boldsymbol{\mu}}_2(s)$ are parallel, as well. Therefore, by definition, the centrode $\boldsymbol{d}(s)$ is a framed rectifying curve.

Conversely, let $\gamma(s)$ be a framed rectifying curve in $\mathbb{R}^3$. From Theorem 3, we have:

$$\lambda'(s) = \alpha(s), \quad \lambda(s)p(s) = cq(s) \tag{13}$$

for some constant c.

Let $f(s) = \frac{1}{c}\int_{s_0}^{s} p(u)du$. There exists a framed curve $\boldsymbol{\beta}(t)$ whose framed curvature satisfies $p_\beta(t) = c$ and $q_\beta(t) = \lambda(t)$.

Let us consider the centrode of β, which is given by $\boldsymbol{d}_\beta(t) = \lambda(t)\boldsymbol{\nu}_\beta(t) + c\overline{\boldsymbol{\mu}}_{\beta,2}(t)$, and its reparametrization $\boldsymbol{\chi}(s) = \boldsymbol{d}_\beta(f(s))$. Then:

$$\boldsymbol{\chi}(s) = \lambda(f(s))\boldsymbol{\nu}_\beta(f(s)) + c\overline{\boldsymbol{\mu}}_{\beta,2}(f(s)).$$

This means that $\boldsymbol{\chi}'(s) = \alpha(s)\boldsymbol{\nu}_\beta(f(s))$; thus, $\boldsymbol{\nu}_\chi(s) = \boldsymbol{\nu}_\beta(f(s))$. Differentiating twice, the framed curvature functions of χ are given by $\alpha_\chi(s) = \alpha(s)$, $p_\chi(s) = p_\beta(s)f'(s) = p(s)$ and $q_\chi(s) = q_\beta(s)f'(s) = q(s)$.

Therefore, the framed curves $\gamma(s)$ and $\chi(s)$ have the same framed curvature functions. From the existence theorem and the uniqueness theorem, it follows that χ is congruent to γ. Consequently, the framed rectifying curve γ is the centrode of a framed curve with nonconstant framed curvature $q(s)$ and nonzero constant framed curvature p. □

The framed curve in Theorem 7 can be replaced by a framed curve with nonzero constant framed curvature q and nonconstant framed curvature $p(s)$. In fact, we also have the following theorem:

Theorem 8. *The centrode of a framed curve with nonzero constant framed curvature function $q(s)$ and nonconstant framed curvature function $p(s)$ is a framed rectifying curve. Conversely, one framed rectifying curve in $\mathbb{R}^3$ is the centrode of some framed curve with nonconstant framed curvature function $p(s)$ and nonzero constant framed curvature function $q(s)$.*

The proof can be given in as similar way as Theorem 7.

Remark 3. *The centrode of a framed curve with nonzero constant framed curvature function $p(s)$ and nonzero constant framed curvature function $q(s)$ is a point.*

7. Contact between Framed Rectifying Curves

In this section, the contact between framed rectifying curves is considered. We now introduce the notion of circular rectifying curves as follows.

Definition 5. *Let $\gamma(s)$ be a framed rectifying curve and:*

$$\gamma(s) = \rho(\tan^2(\int |\boldsymbol{g}'(s)|ds + C) + 1)^{\frac{1}{2}}\boldsymbol{g}(s),$$

where ρ is a positive number and C is a constant. We call γ a circular rectifying curve if $\boldsymbol{g}(s)$ is a circle on S^2.

Let $(\gamma, \boldsymbol{\mu}_1, \boldsymbol{\mu}_2) : I \to S^2 \times \Delta_2$ be a framed spherical curve. We choose $\boldsymbol{\mu}_1 = \gamma$, then $\boldsymbol{\nu} = \gamma \times \boldsymbol{\mu}_2$ and $\gamma'(s) = \alpha(s)\boldsymbol{\nu}(s)$. We show that the spherical Frenet–Serret formula of γ is as follows:

$$\begin{cases} \gamma'(s) = \alpha(s)\boldsymbol{\nu}(s) \\ \boldsymbol{\mu}_2'(s) = l(s)\boldsymbol{\nu}(s) \\ \boldsymbol{\nu}'(s) = -\alpha(s)\gamma(s) - l(s)\boldsymbol{\mu}_2(s), \end{cases}$$

where $\langle \boldsymbol{\mu}_2'(s), \boldsymbol{\nu}(s)\rangle = l(s)$. By the curvature functions $\alpha(s)$ and $l(s)$, we show the following proposition for framed spherical curves:

Proposition 2. *Let* $(\gamma, \gamma, \mu_2) : I \to S^2 \times \Delta_2$ *be a framed spherical curve, then* γ *is a circle if and only if* $\alpha(s) \neq 0$ *and* $l(s)/\alpha(s) = constant$*.*

Proof. If $\alpha(s) \neq 0$ and $(l/\alpha)(s) = k$, where k is a constant, then we consider a normal vector field $N(s) = \frac{k^2}{k^2+1}\gamma(s) - \frac{k}{k^2+1}\mu_2(s)$. By taking the derivative of $N(s)$, we have $N'(s) = \frac{k^2}{k^2+1}(\alpha(s)\nu(s) - \alpha(s)\nu(s)) = \mathbf{0}$. This means that $N(s)$ is a constant vector. Moreover, we have:

$$\langle N(s), \gamma(s) - N(s)\rangle = \langle \frac{k^2}{k^2+1}\gamma(s) - \frac{k}{k^2+1}\mu_2(s), \frac{1}{k^2+1}\gamma(s) + \frac{k}{k^2+1}\mu_2(s)\rangle = 0.$$

This means that γ is the intersection of a plane and S^2, so γ is a circle.

Let γ be a circle on S^2. Obviously, γ is a plane curve and $\alpha(s) \neq 0$, so that $\langle \gamma'(s), \gamma''(s) \times \gamma'''(s)\rangle = 0$. Then, we can calculate that $\langle \gamma'(s), \gamma''(s) \times \gamma'''(s)\rangle = \alpha^4(s)l'(s) - \alpha^3(s)\alpha'(s)l(s)$. Since $\alpha(s) \neq 0$, we have $\alpha(s)l'(s) - \alpha'(s)l(s) = 0$. This is equivalent to $(l/\alpha)'(s) = 0$. Thus, $l(s)/\alpha(s) = constant$. □

As a corollary of Proposition 2, we have the following result:

Corollary 1. *Let* $(\gamma, \gamma, \mu_2) : I \to S^2 \times \Delta_2$ *be a framed spherical curve, then* γ *is a great circle on* S^2 *if and only if* $\alpha(s) \neq 0$ *and* $l(s) = 0$*.*

Now, we review the notions of contact between framed curves [14]. Let $(\gamma, \mu_1, \mu_2) : I \to \mathbb{R}^3 \times \Delta_2$; $s \to (\gamma(s), \mu_1(s), \mu_2(s))$ and $(\widetilde{\gamma}, \widetilde{\mu}_1, \widetilde{\mu}_2) : \widetilde{I} \to \mathbb{R}^3 \times \Delta_2$; $u \to (\widetilde{\gamma}(u), \widetilde{\mu}_1(u), \widetilde{\mu}_2(u))$ be framed curves. We say that (γ, μ_1, μ_2) and $(\widetilde{\gamma}, \widetilde{\mu}_1, \widetilde{\mu}_2)$ have k^{th} order contact at $s = s_0$, $u = u_0$ if:

$$(\gamma, \mu_1, \mu_2)(s_0) = (\widetilde{\gamma}, \widetilde{\mu}_1, \widetilde{\mu}_2)(u_0), \frac{d}{ds}(\gamma, \mu_1, \mu_2)(s_0) = \frac{d}{du}(\widetilde{\gamma}, \widetilde{\mu}_1, \widetilde{\mu}_2)(u_0), \ldots,$$
$$\frac{d^{k-1}}{ds^{k-1}}(\gamma, \mu_1, \mu_2)(s_0) = \frac{d^{k-1}}{du^{k-1}}(\widetilde{\gamma}, \widetilde{\mu}_1, \widetilde{\mu}_2)(u_0), \frac{d^k}{ds^k}(\gamma, \mu_1, \mu_2)(s_0) \neq \frac{d^k}{du^k}(\widetilde{\gamma}, \widetilde{\mu}_1, \widetilde{\mu}_2)(u_0).$$

In addition, we say that (γ, μ_1, μ_2) and $(\widetilde{\gamma}, \widetilde{\mu}_1, \widetilde{\mu}_2)$ have at least k^{th} order contact at $s = s_0$, $u = u_0$ if:

$$(\gamma, \mu_1, \mu_2)(s_0) = (\widetilde{\gamma}, \widetilde{\mu}_1, \widetilde{\mu}_2)(u_0), \frac{d}{ds}(\gamma, \mu_1, \mu_2)(s_0) = \frac{d}{du}(\widetilde{\gamma}, \widetilde{\mu}_1, \widetilde{\mu}_2)(u_0), \ldots,$$
$$\frac{d^{k-1}}{ds^{k-1}}(\gamma, \mu_1, \mu_2)(s_0) = \frac{d^{k-1}}{du^{k-1}}(\widetilde{\gamma}, \widetilde{\mu}_1, \widetilde{\mu}_2)(u_0).$$

We generally say that (γ, μ_1, μ_2) and $(\widetilde{\gamma}, \widetilde{\mu}_1, \widetilde{\mu}_2)$ have at least first order contact at any point $s = s_0$, $u = u_0$, up to congruence as framed curves. As a conclusion of Theorem 3.7 in [14], we show the following proposition:

Proposition 3. *Let* $(\gamma, \gamma, \mu_2) : I \to S^2 \times \Delta_2$, $s \to (\gamma(s), \gamma(s), \mu_2(s))$ *and* $(\widetilde{\gamma}, \widetilde{\gamma}, \widetilde{\mu}_2) : \widetilde{I} \to S^2 \times \Delta_2$, $u \to (\widetilde{\gamma}(u), \widetilde{\gamma}(u), \widetilde{\mu}_2(u))$ *be framed spherical curves. If* (γ, γ, μ_2) *and* $(\widetilde{\gamma}, \widetilde{\gamma}, \widetilde{\mu}_2)$ *have at least* $(k+1)^{th}$ *order contact at* $s = s_0$, $u = u_0$*, we have:*

$$\alpha(s_0) = \widetilde{\alpha}(u_0), \frac{d}{ds}\alpha(s_0) = \frac{d}{du}\widetilde{\alpha}(u_0), \ldots, \frac{d^{k-1}}{ds^{k-1}}\alpha(s_0) = \frac{d^{k-1}}{du^{k-1}}\widetilde{\alpha}(u_0), \tag{14}$$

$$l(s_0) = \widetilde{l}(u_0), \frac{d}{ds}l(s_0) = \frac{d}{du}\widetilde{l}(u_0), \ldots, \frac{d^{k-1}}{ds^{k-1}}l(s_0) = \frac{d^{k-1}}{du^{k-1}}\widetilde{l}(u_0). \tag{15}$$

Conversely, if the conditions (14) and (15) hold, then (γ, γ, μ_2) *and* $(\widetilde{\gamma}, \widetilde{\gamma}, \widetilde{\mu}_2)$ *have at least* $(k+1)^{th}$ *order contact at* $s = s_0$, $u = u_0$*, up to congruence as framed spherical curves.*

Now, we consider the contact between circles and framed spherical curves. We have a corollary of Propositions 2 and 3 as follows:

Corollary 2. *Let $(\gamma, \gamma, \mu_2) : I \to S^2 \times \Delta_2$ be a framed spherical curve. γ and a circle have at least $(k+1)^{th}$ order contact at $s = s_0$ if and only if there exists a constant σ such that:*

$$l(s_0) = \sigma\alpha(s_0), \frac{d}{ds}l(s_0) = \sigma\frac{d}{ds}\alpha(s_0), \dots, \frac{d^{k-1}}{ds^{k-1}}l(s_0) = \sigma\frac{d^{k-1}}{ds^{k-1}}\alpha(s_0).$$

For the construction of the framed rectifying curve in Theorem 4, we fix positive number ρ and constant C. Let $g_i : I \to S^2$ $(i = 1, 2)$ be framed spherical curves. We know γ_1, γ_2 have k^{th} order contact at s_0 if and only if g_1, g_2 have k^{th} order contact at s_0. By Corollary 2, we have the following theorem, which can describe the contact between framed rectifying curves and circular rectifying curves.

Theorem 9. *Let γ be a framed rectifying curve and $\alpha(s)$ and $l(s)$ be curvature functions of the corresponding framed spherical curve. Then, γ and a circular rectifying curve have at least k^{th} order $(k \geq 2)$ contact at s_0 if and only if there exists a constant σ such that:*

$$l(s_0) = \sigma\alpha(s_0), \frac{d}{ds}l(s_0) = \sigma\frac{d}{ds}\alpha(s_0), \dots, \frac{d^{k-2}}{ds^{k-2}}l(s_0) = \sigma\frac{d^{k-2}}{ds^{k-2}}\alpha(s_0).$$

Author Contributions: Conceptualization, Y.W.; Writing—Original Draft Preparation, Y.W.; Calculations, D.P.; Manuscript Correction, D.P.; Giving the Examples, R.G.; Drawing the Pictures, R.G.

Funding: This research was funded by National Natural Science Foundation of China grant numbers 11271063 and 11671070, the Fundamental Research Funds for the Central Universities grant number 3132018220, and the Project of Science and Technology of Jilin Provincial Education Department grant number JJKH2090547KJ.

Acknowledgments: This work was partially supported by the National Natural Science Foundation of China Nos. 11271063 and 11671070, the Fundamental Research Funds for the Central Universities No. 3132018220, and The Project of Science and Technology of Jilin Provincial Education Department No. JJKH2090547KJ.

Conflicts of Interest: The authors declare no conflict of interest.

References

1. Mikula, K.; Sevcovic, D. Evolution of curves on a surface driven by the geodesic curvature and external force. *Appl. Anal.* **2006**, *85*, 345–362. [CrossRef]
2. Izumiya, S.; Takeuchi, N. Generic properties of helices and Bertrand curves. *J. Geom.* **2002**, *74*, 97–109. [CrossRef]
3. Liu, Y.; Ru, M. Degeneracy of holomorphic curves in surfaces. *Sci. China Ser. A* **2005**, *48*, 156–167. [CrossRef]
4. Chen, B.Y. When does the position vector of a space curve always lie in its rectifying plane? *Am. Math. Mon.* **2003**, *110*, 147–152. [CrossRef]
5. Chen, B.Y.; Dillen, F. Rectifying curves as centrodes and extremal curves. *Bull. Inst. Math. Acad. Sin.* **2005**, *33*, 77–90.
6. Yeh, H.; Abrams, J.I. *Principles of Mechanics of Solids and Fluids*; McGraw-Hill: New York, NY, USA, 1960; Volume 1.
7. Ogston, N.; King, G.; Gertzbein, S.; Tile, M.; Kapasouri, A.; Rubenstein, J. Centrode patterns in the lumbar spine-base-line studies in normal subjects. *Spine* **1986**, *11*, 591–595. [CrossRef] [PubMed]
8. Weiler, P.J.; Bogoch, R.E. Kinematics of the distal radioulnar joint in rheumatoid-arthritis-an in-vivo study using centrode analysis. *J. Hand Surg.* **1995**, *20*, 937–943. [CrossRef]
9. Gertzbein, S.; Seligman, J.; Holtby, R.; Chan, K.; Ogston, N.; Kapasouri, A.; Tile, M.; Cruickshank, B. Centrode patterns and segmental instability in degenerative disk disease. *Spine* **1985**, *10*, 257–261. [CrossRef] [PubMed]
10. Ilarslan, K.; Nesovic, E. On rectifying curves as centrodes and extremal curves in the Minkowski 3-space. *Novi Sad J. Math.* **2007**, *37*, 53–64.
11. Ilarslan, K.; Nesovic, E. Some characterizations of null, pseudo null and partially null rectifying curves in Minkowski space-time. *Taiwan. J. Math.* **2008**, *12*, 1035–1044.

12. Πarslan, K.; Nesovic, E. Some characterizations of rectifying curves in the Euclidean space E4. *Turk. J. Math.* **2008**, *32*, 21–30.
13. Grbovic, M.; Nesovic, E. Some relations between rectifying and normal curves in Minkowski 3-space. *Math. Commun.* **2012**, *17*, 655–664.
14. Honda, S.; Takahashi, M. Framed curves in the Euclidean space. *Adv. Geom.* **2016**, *16*, 265–276. [CrossRef]
15. Chen, L.; Takahashi, M. Dualities and evolutes of fronts in hyperbolic and de Sitter space. *J. Math. Anal. Appl.* **2016**, *437*, 133–159. [CrossRef]
16. Bishop, R.L. There is more than one way to frame a curve. *Am. Math. Mon.* **1975**, *82*, 246–251. [CrossRef]

mathematics

MDPI

Article

Hypersurfaces with Generalized 1-Type Gauss Maps

Dae Won Yoon [1], Dong-Soo Kim [2], Young Ho Kim [3] and Jae Won Lee [1,*]

1 Department of Mathematics Education and RINS, Gyeongsang National University, Jinju 52828, Korea; dwyoon@gnu.ac.kr
2 Department of Mathematics, Chonnam National University, Gwangju 61186, Korea; dosokim@chonnam.ac.kr
3 Department of Mathematics, Kyungpook National University, Daegu 41566, Korea; yhkim@knu.ac.kr
* Correspondence: leejaew@gnu.ac.kr; Tel.: +82-55-772-2251

Received: 18 May 2018; Accepted: 23 July 2018; Published: 26 July 2018

Abstract: In this paper, we study submanifolds in a Euclidean space with a generalized 1-type Gauss map. The Gauss map, G, of a submanifold in the n-dimensional Euclidean space, $\mathbb{E}^n$, is said to be of generalized 1-type if, for the Laplace operator, Δ, on the submanifold, it satisfies $\Delta G = fG + gC$, where C is a constant vector and f and g are some functions. The notion of a generalized 1-type Gauss map is a generalization of both a 1-type Gauss map and a pointwise 1-type Gauss map. With the new definition, first of all, we classify conical surfaces with a generalized 1-type Gauss map in $\mathbb{E}^3$. Second, we show that the Gauss map of any cylindrical surface in $\mathbb{E}^3$ is of the generalized 1-type. Third, we prove that there are no tangent developable surfaces with generalized 1-type Gauss maps in $\mathbb{E}^3$, except planes. Finally, we show that cylindrical hypersurfaces in $\mathbb{E}^{n+2}$ always have generalized 1-type Gauss maps.

Keywords: conical surface; developable surface; generalized 1-type Gauss map; cylindrical hypersurface

1. Introduction

The notion of finite type submanifolds in a Euclidean space or a pseudo-Euclidean space was introduced by Chen in the 1980s [1]. He also extended this notion to a general differential map, namely, the Gauss map, on the submanifolds. The notions of finite type immersion and finite type Gauss map are useful tools for investigating and characterizing many important submanifolds [1–12]. Moreover, Chen et al. dealed with the finite type Gauss map as an immersion and with its relation to the topology of some submanifolds [13,14].

The simplest type of finite type Gauss map is the 1-type. A submanifold, M, of a Euclidean space or a pseudo-Euclidean space has a 1-type Gauss map if the Gauss map, G, of M satisfies

$$\Delta G = \lambda(G + C) \tag{1}$$

for some $\lambda \in \mathbb{R}$ and has a constant vector, C, where Δ denotes the Laplace operator defined on M. Planes, circular cylinders and spheres in $\mathbb{E}^3$ are typical examples of surfaces with 1-type Gauss maps.

As a generalization of a 1-type Gauss map, the first and third authors introduced the notion of a pointwise 1-type Gauss map of submanifolds in reference [15]. A submanifold is said to have a *pointwise 1-type Gauss map* if the Laplacian of its Gauss map, G, takes the form

$$\Delta G = f(G + C) \tag{2}$$

for a non-zero smooth function, f, and a constant vector, C. More precisely, a pointwise 1-type Gauss map is said to be of the first kind if $C = 0$ in (2); otherwise, it is said to be of the second kind. A helicoid, a catenoid and a right cone in $\mathbb{E}^3$ are typical examples of surfaces with pointwise 1-type Gauss maps.

Many results of submanifolds with pointwise 1-type Gauss maps in ambient spaces were obtained in references [6,16–27]. On the other hand, it is well-known that a circular cylinder in $\mathbb{E}^3$ has a usual 1-type Gauss map. However, we consider the following cylindrical surface parameterized by

$$x(s,t) = \left(\frac{s}{2}\cos(\ln s) + \frac{s}{2}\sin(\ln s), -\frac{s}{2}\cos(\ln s) + \frac{s}{2}\sin(\ln s), t\right).$$

Then, the Gauss map, G, of the surface is given by

$$G = (-\sin(\ln s), \cos(\ln s), 0).$$

We can easily show that the Gauss map, G, satisfies

$$\Delta G = \frac{1}{s^2}\left(1 + \cot(\ln s)\right) G - \frac{1}{s^2}\csc(\ln s)(0,1,0),$$

which yields a Gauss map, G, that is neither of usual 1-type, nor of pointwise 1-type.

In this reason, we have the following definition:

Definition 1. *A submanifold, M, of a Euclidean space is said to have a generalized 1-type Gauss map if the Gauss map, G, on M satisfies the equation*

$$\Delta G = fG + gC \tag{3}$$

for some smooth functions (f, g) and has a constant vector, C.

If both f and g are constant in (3), then M has a 1-type Gauss map. If $f = g$ in (3), then M has a pointwise 1-type Gauss map. Hence, the notion of a generalized 1-type Gauss map is a generalization of both a 1-type Gauss map and a pointwise 1-type Gauss map.

In [22], Dursun studied flat surfaces in $\mathbb{E}^3$ with a pointwise 1-type Gauss map and proved the following proposition.

Proposition 1. *Let M be a flat surface in $\mathbb{E}^3$. Then, M has a pointwise 1-type Gauss map of the second kind if and only if M is an open part of one of the following surfaces:*

(1) *A plane in $\mathbb{E}^3$,*
(2) *A right circular cone in $\mathbb{E}^3$,*
(3) *A cylinder, up to a rigid motion, parameterized by*

$$x(s,t) = \gamma(s) + t\beta,$$

where $\gamma = \gamma(s)$ is a unit speed planar base curve with curvature $k = k(s)$ satisfying the ordinary differential equation

$$(\frac{dk}{ds})^2 = k^4(s)\{ak^2(s) + 2bk(s) - 1\}$$

for some real numbers, a and $b(\neq 0)$, and the director vector $\beta = (0,0,1)$.

In this paper, we study developable surfaces in $\mathbb{E}^3$: cylindrical surfaces, conical surfaces and tangent developable surfaces. In Section 3, we completely classify developable surfaces with generalized a 1-type Gauss map and give some examples. In the last section, we prove that cylindrical hypersurfaces in $\mathbb{E}^{n+2}$ always have generalized 1-type Gauss maps.

Throughout this paper, we assume that all objects are smooth and all surfaces are connected unless mentioned.

2. Preliminaries

Let $x : M \longrightarrow \mathbb{E}^m$ be an isometric immersion from an n-dimensional Riemannian manifold, M, into $\mathbb{E}^m$. Denote the Levi–Civita connections of M and $\mathbb{E}^m$ by ∇ and $\widetilde{\nabla}$, respectively. Let X and Y be vector fields tangent to M, and let ξ be a unit normal vector field of M. Then, the Gauss and Weingarten formulas are given by

$$\widetilde{\nabla}_X Y = \nabla_X Y + h(X, Y), \tag{4}$$

$$\widetilde{\nabla}_X \xi = -A_\xi X + D_X \xi, \tag{5}$$

respectively. Here, h is the second fundamental form; D is the normal connection defined on the normal bundle; and A_ξ is the shape operator (or the Weingarten operator) in the direction of ξ on M. Note that the second fundamental form, h, and the shape operator, A_ξ, are related by

$$\langle h(X, Y), \xi \rangle = \langle A_\xi X, Y \rangle. \tag{6}$$

The mean curvature vector field, $\overrightarrow{H}$, is defined by

$$\overrightarrow{H} = \frac{1}{n} \mathrm{tr} h, \tag{7}$$

where tr$h$ is the trace of h. The mean curvature, H, of M is given by $H = \sqrt{\langle \overrightarrow{H}, \overrightarrow{H} \rangle}$.

Moreover, the Laplace operator, Δ, acting on a scalar valued function, ϕ, is given by

$$\Delta \phi = -\sum_{i=1}^{n} (\widetilde{\nabla}_{e_i} \widetilde{\nabla}_{e_i} \phi - \widetilde{\nabla}_{\nabla_{e_i} e_i} \phi), \tag{8}$$

where $\{e_1, ..., e_n\}$ is an orthonormal local tangent frame on M. Or, locally, it is expressed as

$$\Delta \phi = -\frac{1}{\sqrt{g}} \sum_{i,j=1}^{n} \frac{\partial}{\partial x_i} (\sqrt{g} g^{ij} \frac{\partial \phi}{\partial x_j}), \tag{9}$$

where (g^{ij}) and g denote the inverse matrix and the determinant of the matrix (g_{ij}), respectively, with the coefficients g_{ij} of the Riemannian metric $\langle \cdot, \cdot \rangle$ on M induced from that of $\mathbb{E}^m$.

3. Surfaces with Generalized 1-Type Gauss Maps

In this section, we completely classify developable surfaces in $\mathbb{E}^3$ with a generalized 1-type Gauss map.

A regular surface in $\mathbb{E}^3$ whose Gaussian curvature vanishes is called a *developable surface*, whose surface is a cylindrical surface, a conical surface or a tangent developable surface [28].

For a hypersurface in a Euclidean space, the next lemma plays an important role in our paper [21].

Lemma 1. *Let M be a hypersurface of $\mathbb{E}^{n+2}$. Then, the Laplacian of the Gauss map, G, is given by*

$$\Delta G = ||A_G||^2 G + (n+1) \nabla H, \tag{10}$$

where ∇H is the gradient of the mean curvature, H; A_G is the shape operator of M; and $||A_G||^2 = tr(A_G^2)$.

Suppose that a developable surface in $\mathbb{E}^3$ has a generalized 1-type Gauss map, that is, the Gauss map G of the surface satisfies the condition

$$\Delta G = fG + gC \tag{11}$$

for some smooth functions, f, g, and a constant vector, C. It follows from (10) that M has generalized 1-type Gauss map with $C = 0$, that is, M has a pointwise 1-type Gauss map of the first kind if and only if M has a constant mean curvature, H. If f and g are equal to each other with $C \neq 0$, then M has a pointwise 1-type Gauss map of the second kind and the results occur in [22]. Therefore, sometimes, in the proof of this paper, we assume that $f \neq g$ has non-zero functions and $C \neq 0$.

By combining (10) and (11) and taking the inner product with the orthonormal local frame e_1, e_2 and G, respectively, we have

$$\begin{aligned} 2e_1(H) &= gC_1, \\ 2e_2(H) &= gC_2, \\ ||A_G||^2 &= f + gC_3, \end{aligned} \tag{12}$$

where $C = C_1e_1 + C_2e_2 + C_3G$ with $C_1 = \langle C, e_1\rangle$, $C_2 = \langle C, e_2\rangle$ and $C_3 = \langle C, G\rangle$.

3.1. Conical Surfaces

A conical surface, M, in $\mathbb{E}^3$ can be parametrized by

$$x(s,t) = \alpha_0 + t\beta(s), \quad s \in I, \quad t > 0,$$

such that $\langle \beta(s), \beta(s)\rangle = \langle \beta'(s), \beta'(s)\rangle = 1$, where α_0 is a constant vector. We take the orthonormal tangent frame, $\{e_1, e_2\}$, on M such that $e_1 = \frac{1}{t}\frac{\partial}{\partial s}$ and $e_2 = \frac{\partial}{\partial t}$. The Gauss map of M is given by $G = e_1 \times e_2$. Through a direct calculation, we have

$$\begin{aligned} \widetilde{\nabla}_{e_1}e_1 &= -\frac{1}{t}e_2 - \frac{\kappa_g(s)}{t}G, \quad \widetilde{\nabla}_{e_1}e_2 = \frac{1}{t}e_1, \\ \widetilde{\nabla}_{e_2}e_1 &= \widetilde{\nabla}_{e_2}e_2 = 0, \quad \widetilde{\nabla}_{e_1}G = \frac{\kappa_g}{t}e_1, \quad \widetilde{\nabla}_{e_2}G = 0, \end{aligned} \tag{13}$$

where $\kappa_g(s) = \langle \beta(s), \beta'(s) \times \beta''(s)\rangle$ denotes the geodesic curvature of β in the unit sphere, $\mathbb{S}^2(1)$. We may assume that $\kappa_g(s) \neq 0, s \in I$; otherwise, the conical surface is an open part of a plane. Furthermore, by reversing the orientation of the spherical curve, β, we may assume that the geodesic curvature, κ_g, of β is positive. It follows from (13) that the mean curvature, H, and the trace, $||A_G||^2$, of the square of the shape operator are given by

$$H = -\frac{\kappa_g(s)}{2t}, \quad ||A_G||^2 = \frac{\kappa_g^2(s)}{t^2}. \tag{14}$$

Suppose that M has a generalized 1-type Gauss map, that is, the Gauss map, G, of the conical surface satisfies (11). Then, since $C_1 = \langle C, \beta'\rangle$, $C_2 = \langle C, \beta\rangle$ and $C_3 = \langle C, \beta' \times \beta\rangle$, the components $C_i (i = 1, 2, 3)$ of the constant vector, C, are functions of only s. Let us differentiate C_1, C_2 and C_3 with respect to e_1. Then, from (13), we have the following:

$$C_1'(s) + C_2(s) + \kappa_g(s)C_3(s) = 0, \tag{15}$$

$$C_2'(s) - C_1(s) = 0, \tag{16}$$

$$C_3'(s) - \kappa_g(s)C_1(s) = 0. \tag{17}$$

With the help of (14), (12) can be written as

$$-\frac{1}{t^2}\kappa_g'(s) = gC_1, \tag{18}$$

$$\frac{\kappa_g(s)}{t^2} = gC_2, \tag{19}$$

$$\frac{\kappa_g^2(s)}{t^2} = f + gC_3. \tag{20}$$

By combining (18) and (19) and using (16), we have

$$g\left(\kappa_g(s)C_2\right)' = 0.$$

Since $g \neq 0$, $\kappa_g(s)C_2$ is a non-zero constant, say c, we obtain

$$C_2 = \frac{c}{\kappa_g(s)}. \tag{21}$$

Together with (19), this implies

$$g = \frac{\kappa_g^2(s)}{ct^2}, \tag{22}$$

and hence, from (18), we get

$$C_1 = -\frac{c\kappa_g'(s)}{\kappa_g^2(s)}. \tag{23}$$

Thus, it follows from (15) that we have

$$C_3 = \frac{c\left(\kappa_g(s)\kappa_g''(s) - 2\kappa_g'(s)^2 - \kappa_g^2(s)\right)}{\kappa_g^4(s)}. \tag{24}$$

Note that the function f is determined by (20), (22) and (24).

Now, we have $\varphi(s) = 1/\kappa_g(s) > 0$. Then, (21) and (23) become, respectively,

$$C_2 = c\varphi \tag{25}$$

and

$$C_1 = c\varphi'. \tag{26}$$

Furthermore, it follows from (17) and (24) that

$$C_3 = -c(\varphi\varphi'' + \varphi^2) \tag{27}$$

and

$$C_3' = c\frac{\varphi'}{\varphi}. \tag{28}$$

Thus, from (27) and (28) we see that the function φ must satisfy the following nonlinear differential equation of order 3:

$$\varphi^2\varphi''' + \varphi\varphi'\varphi'' + 2\varphi^2\varphi' + \varphi' = 0. \tag{29}$$

In order to solve (29), first, we put $p = d\varphi/ds$. Then the differential equation (29) becomes

$$p\left(\varphi^2 p\frac{d^2p}{d\varphi^2} + \varphi^2(\frac{dp}{d\varphi})^2 + \varphi p\frac{dp}{d\varphi} + 2\varphi^2 + 1\right) = 0,$$

which can be rewritten as

$$\varphi p\left(\frac{d}{d\varphi}(\varphi p\frac{dp}{d\varphi}) + 2\varphi + \frac{1}{\varphi}\right) = 0. \tag{30}$$

Since $\varphi > 0$, we divide into two cases, as follows.

Case 1. $p = d\varphi/ds = 0$. The geodesic curvature, κ_g, is a nonzero constant, that is, the spherical curve, $\beta(s)$, is a small circle. Therefore, M is an open part of a right circular cone, and M has a pointwise 1-type Gauss map.

Case 2. $p = d\varphi/ds \neq 0$.

From (30), we obtain

$$\frac{d}{d\varphi}(\varphi p\frac{dp}{d\varphi}) + 2\varphi + \frac{1}{\varphi} = 0, \tag{31}$$

which yields

$$\varphi p\frac{dp}{d\varphi} + \varphi^2 + \ln\varphi = \frac{a}{2} \tag{32}$$

for some constant, a. By integrating (32), we have

$$p^2 = a\ln\varphi + b - \varphi^2 - (\ln\varphi)^2 \tag{33}$$

for some constant, b. Recalling $p = d\varphi/ds$, from (33), one gets

$$\frac{d\varphi}{ds} = \pm\left(a\ln\varphi + b - \varphi^2 - (\ln\varphi)^2\right)^{\frac{1}{2}}, \tag{34}$$

which is equivalent to

$$\frac{d\varphi}{(a\ln\varphi + b - \varphi^2 - (\ln\varphi)^2)^{\frac{1}{2}}} = \pm ds. \tag{35}$$

Hence, for an indefinite integral, $F(t)$, of the function $\psi(t) = (a\ln t + b - t^2 - (\ln t)^2)^{-1/2}$, we see that

$$F(\varphi) = \pm s, \tag{36}$$

where the signature is determined according to whether the derivative of φ is positive or not. Thus we get

$$\kappa_g(s) = \frac{1}{\varphi(s)} = \frac{1}{F^{-1}(\pm s)}. \tag{37}$$

Furthermore, it follows from (25)–(27) that C can be expressed as

$$C = c\left(\varphi' e_1 + \varphi e_2 - (\varphi\varphi'' + \varphi^2)G\right), \tag{38}$$

or equivalently,

$$C = c\left(-\frac{\kappa_g'}{\kappa_g^2}e_1 + \frac{1}{\kappa_g}e_2 + \frac{\kappa_g(s)\kappa_g''(s) - 2\kappa_g'(s)^2 - \kappa_g^2(s)}{\kappa_g^4(s)}G\right). \tag{39}$$

Conversely, for some constants, a and b, such that the function

$$\psi(t) = \left(a\ln t + b - t^2 - (\ln t)^2\right)^{-1/2} \tag{40}$$

is well-defined on some interval, $J \subset (0, \infty)$, we take an indefinite integral, $F(t)$, of the function $\psi(t)$. If we denote the image of the function, F, by I, then $F : J \to I$ is a strictly increasing function with $F'(t) = \psi(t)$. Let us consider the function $\varphi = \varphi_\pm$, defined by $\varphi_\pm(s) = F^{-1}(\pm s)$, which maps the interval, $\pm I$, onto J, respectively. Here $-I$ means the interval $\{-s | s \in I\}$. Then, the function $\varphi = \varphi_\pm$ is positive for the interval $I_\pm$ (say, I) and satisfies $F(\varphi) = \pm s$.

For any unit speed spherical curve $\beta(s)$ in the unit sphere $\mathbb{S}^2(1)$ with the geodesic curvature $\kappa_g(s) = 1/\varphi(s)$, we consider a surface M in $\mathbb{E}^3$ to be parametrized by

$$x(s,t) = \alpha_0 + t\beta(s), \quad s \in I, \quad t > 0, \tag{41}$$

where α_0 is a constant vector. Given any non-zero constant, c, we put

$$f(s,t) = \frac{1}{t^2\varphi^2(s)}\left(\varphi(s)\varphi''(s) + \varphi^2(s) + 1\right), \quad g(s,t) = \frac{1}{ct^2\varphi^2(s)}. \tag{42}$$

For the orthonormal tangent frame, $\{e_1, e_2\}$, on M, such that $e_1 = \frac{1}{t}\frac{\partial}{\partial s}$ and $e_2 = \frac{\partial}{\partial t}$ and the Gauss map of M given by $G = e_1 \times e_2$, we put

$$C = c\{\varphi'(s)e_1 + \varphi(s)e_2 - \left(\varphi(s)\varphi''(s) + \varphi^2(s)\right)G\}. \tag{43}$$

Note that it follows from the definition of φ that the function φ satisfies (29). Hence, by using (13), it is straightforward to show that

$$\widetilde{\nabla}_{e_1}C = \widetilde{\nabla}_{e_2}C = 0, \tag{44}$$

which implies that C is a constant vector. Furthermore, similar to the first part of this subsection, the Gauss map, G, of the conical surface, M, satisfies

$$\Delta G = fG + gC,$$

where f, g and C are given in (42) and (43), respectively. This shows that M has a generalized 1-type Gauss map.

Thus, we have the following theorem 1:

Theorem 1. *A conical surface in $\mathbb{E}^3$ has a generalized 1-type Gauss map if and only if it is an open part of one of the following surfaces:*

(1) A plane,
(2) A right circular cone,
(3) A conical surface parameterized by

$$x(s,t) = \alpha_0 + t\beta(s),$$

where α_0 is a constant vector and $\beta(s)$ is a unit speed spherical curve in the unit sphere $\mathbb{S}^2(1)$ with a positive geodesic curvature, κ_g, which for some indefinite integral $F(t)$ of the function $\psi(t) = \left(a\ln t + b - t^2 - (\ln t)^2\right)^{-1/2}$ with $a, b \in R$, is given by

$$\kappa_g(s) = \frac{1}{F^{-1}(\pm s)}.$$

3.2. Cylindrical Surfaces

In this subsection, we prove the following theorem:

Theorem 2. *All cylindrical surfaces in $\mathbb{E}^3$ have a generalized 1-type Gauss map.*

Proof. Let M be a cylindrical surface in $\mathbb{E}^3$ generated by a base curve, $\alpha(s)$, and a constant vector, β. Then, M can be parametrized by

$$x(s,t) = \alpha(s) + t\beta,$$

such that $\langle\alpha'(s), \alpha'(s)\rangle = 1$, $\langle\alpha'(s), \beta\rangle = 0$ and $\langle\beta, \beta\rangle = 1$. Hence, the base curve, $\alpha(s)$, is a unit speed plane curve. Let us denote the curvature function of $\alpha(s)$ by $\kappa(s)$.

Consider an orthonormal frame $\{e_1, e_2\}$ on M such that $e_1 = \frac{\partial}{\partial t}$ and $e_2 = \frac{\partial}{\partial s}$. Then, the Gauss map, G, of M is given by $G = e_1 \times e_2$. By direct calculation, we obtain

$$\begin{aligned} &\tilde{\nabla}_{e_1} e_1 = \tilde{\nabla}_{e_1} e_2 = \tilde{\nabla}_{e_2} e_1 = 0, \quad \tilde{\nabla}_{e_2} e_2 = \kappa(s) G, \\ &\tilde{\nabla}_{e_1} G = 0, \quad \tilde{\nabla}_{e_2} G = -\kappa(s) e_2. \end{aligned} \tag{45}$$

It follows from (45) that the mean curvature, H, and the trace $||A_G||^2$ of the square of the shape operator are given by

$$H = \frac{\kappa(s)}{2}, \quad ||A_G||^2 = \kappa^2(s), \tag{46}$$

which are functions of only s.

First, suppose that M has a generalized 1-type Gauss map. Together with (46), the first equation of (12) shows that $C_1 = 0$. Hence, (12) can be rewritten as

$$\begin{aligned} \kappa^2(s) &= f + gC_3, \\ \kappa'(s) &= gC_2. \end{aligned} \tag{47}$$

Since $C_2 = \langle C, \alpha'(s) \rangle$ and $C_3 = \langle C, \beta \times \alpha'(s) \rangle$, C_2 and C_3 are functions of only s. By differentiating C_2 and C_3 with respect to e_2, the component functions of C satisfy the following equations:

$$\begin{aligned} C_2'(s) - \kappa(s) C_3(s) = 0, \\ C_3'(s) + \kappa(s) C_2(s) = 0, \end{aligned} \tag{48}$$

which yield $C_2^2(s) + C_3^2(s) = c^2$ for some non-zero constant, c. We may put

$$C_2(s) = c \sin \theta(s), \quad C_3(s) = c \cos \theta(s) \tag{49}$$

with $\theta'(s) = \kappa(s)$. Therefore, the constant vector, C, becomes

$$C = c \sin \theta(s) e_2 + c \cos \theta(s) G. \tag{50}$$

By combining (47) and (49), one also gets

$$g = \frac{\kappa'(s)}{c \sin \theta(s)}, \quad f = \kappa^2(s) - \kappa'(s) \cot \theta(s). \tag{51}$$

Conversely, for any cylindrical surface, we choose a curve, $\alpha(s)$, and a unit vector, β, such that the cylindrical surface is parametrized by $x(s,t) = \alpha(s) + t\beta$ with $\langle \alpha'(s), \alpha'(s) \rangle = 1$, $\langle \alpha'(s), \beta \rangle = 0$. Then, for a non-zero constant, c, and an indefinite integral, $\theta(s)$, of the curvature function, $\kappa(s)$, of α, we put

$$C = c \sin \theta(s) e_2 + c \cos \theta(s) G, \tag{52}$$

where $e_1 = \frac{\partial}{\partial t}$, $e_2 = \frac{\partial}{\partial s}$ and $G = e_1 \times e_2$. It follows from (45) that $\tilde{\nabla}_{e_1} C = 0$ and $\tilde{\nabla}_{e_2} C = 0$, which shows that C is a constant vector. Furthermore, it is straightforward to show that the Gauss map, G, of the cylindrical surface satisfies

$$\Delta G = fG + gC,$$

where f, g and C are given in (51) and (52), respectively. This shows that the cylindrical surface has a generalized 1-type Gauss map. □

Example 1. *We consider the surface to be parameterized by*

$$x(s,t) = \left(2\cos(\sqrt{s}) + 2\sqrt{s}\sin(\sqrt{s}), 2\sin(\sqrt{s}) - 2\sqrt{s}\cos(\sqrt{s}), t\right).$$

Then, the surface is cylindrical, generated by the plane curve with the curvature $\kappa(s) = \frac{1}{2\sqrt{s}}$, and its Gauss map G is given by

$$G = \left(\sin(\sqrt{s}), -\cos(\sqrt{s}), 0\right).$$

From this, the Laplacian of G can be expressed as

$$\begin{aligned}\Delta G &= \left(\frac{1}{4s\sqrt{s}}\cos(\sqrt{s}) + \frac{1}{4s}\sin(\sqrt{s}), \frac{1}{4s\sqrt{s}}\sin(\sqrt{s}) - \frac{1}{4s}\cos(\sqrt{s}), 0\right) \\ &= \frac{1}{4s}\left(1 + \frac{\cot(\sqrt{s})}{\sqrt{s}}\right)G + \frac{\csc(\sqrt{s})}{4s\sqrt{s}}C,\end{aligned}$$

where $C = (0, 1, 0)$.

The plane curve and the cylindrical surface in Example 1 are shown in Figures 1 and 2, respectively.

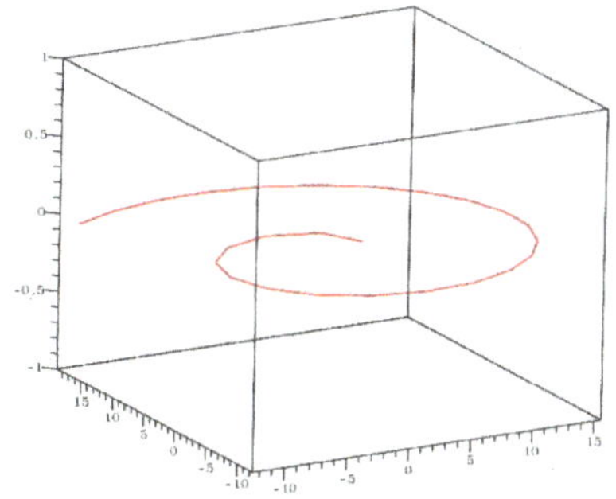

Figure 1. The plane curve in Example 1.

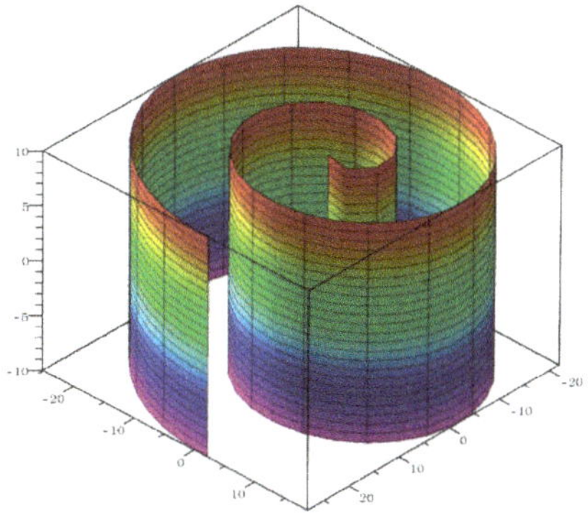

Figure 2. The cylindrical surface in Example 1.

3.3. Tangent Developable Surfaces

In this subsection, we prove the following theorem:

Theorem 3. *A tangent developable surface in $\mathbb{E}^3$ with a generalized 1-type Gauss map is an open part of a plane.*

Proof. Let M be a tangent developable surface in $\mathbb{E}^3$. Then, M is locally parametrized by

$$x(s,t) = \alpha(s) + t\alpha'(s), \quad s \in I, \quad t \neq 0,$$

where $\alpha(s)$ is a unit speed curve with non-zero curvature $\kappa(s)$ in $\mathbb{E}^3$. Let us denote the unit tangent vector, principal normal vector and binormal vector of $\alpha(s)$, by T, N and B, respectively. The natural frames, $\{x_s, x_t\}$ of x, are given by

$$x_s = T + t\kappa(s)N, \quad x_t = \alpha'(s) = T.$$

The parametrization x is regular whenever $t\kappa(s) \neq 0$. We take the orthonormal frame, $\{e_1, e_2\}$, on M such that

$$\begin{aligned} e_1 &= \frac{\partial}{\partial t} = T, \\ e_2 &= \frac{1}{t\kappa(s)}\left(\frac{\partial}{\partial s} - \frac{\partial}{\partial t}\right) = N. \end{aligned} \tag{53}$$

Then, the Gauss map, G, of M is given by $G = e_1 \times e_2 = T \times N = B$. By direct calculation, we obtain

$$\begin{aligned} &\widetilde{\nabla}_{e_1}e_1 = \widetilde{\nabla}_{e_1}e_2 = 0, \quad \widetilde{\nabla}_{e_2}e_1 = \frac{1}{t}e_2, \\ &\widetilde{\nabla}_{e_2}e_2 = -\frac{1}{t}e_1 + \frac{\tau}{t\kappa}G, \widetilde{\nabla}_{e_1}G = 0, \quad \widetilde{\nabla}_{e_2}G = \frac{\tau}{t\kappa}e_2, \end{aligned} \tag{54}$$

which yields

$$H = \frac{\tau}{2t\kappa}, \quad ||A_G||^2 = \left(\frac{\tau}{t\kappa}\right)^2. \tag{55}$$

Now, we suppose that the tangent developable surface, M, has a generalized 1-type Gauss map. Since $C_1 = \langle C, T\rangle$, $C_2 = \langle C, N\rangle$ and $C_3 = \langle C, B\rangle$, the components of C are functions of s only. Hence, it follows from (54) that the components of C satisfy the following:

$$C_1' - \kappa C_2 = 0, \tag{56}$$

$$C_2' + \kappa C_1 - \tau C_3 = 0, \tag{57}$$

$$C_3' + \tau C_2 = 0. \tag{58}$$

Due to (55), (12) can be rewritten as

$$-\frac{\tau}{t^2\kappa} = gC_1, \tag{59}$$

$$\frac{1}{t^2\kappa}\left(\left(\frac{\tau}{\kappa}\right)' + \frac{\tau}{t\kappa}\right) = gC_2, \tag{60}$$

$$\frac{\tau^2}{t^2\kappa^2} = f + gC_3. \tag{61}$$

By combining (59) and (14), one finds that

$$\tau C_2 + \left(\left(\frac{\tau}{\kappa}\right)' + \frac{\tau}{t\kappa}\right)C_1 = 0, \tag{62}$$

or equivalently,

$$\left(\left(\frac{\tau}{\kappa}\right)' C_1 + \tau C_2\right)t\kappa + \tau C_1 = 0. \tag{63}$$

Since the parameter $t(\neq 0)$ is arbitrary, from (63), we have

$$\begin{aligned} \left(\frac{\tau}{\kappa}\right)' C_1 + \tau C_2 &= 0, \\ \tau C_1 &= 0. \end{aligned} \tag{64}$$

Finally, we suppose that the torsion, $\tau(s)$, of the curve, $\alpha(s)$, does not vanish identically. Then, since the set $J = \{s \in I | \tau(s) \neq 0\}$ is non-empty, (64) shows that $C_1 = 0$ and $C_2 = 0$. From this and (57), we have $C_3 = 0$. In the long run, one gets $C = 0$. It follows from (12) and (55) that the mean curvature, $H = \tau(s)/(2t\kappa(s))$, is constant, which shows that τ must vanish identically. That is, $J = \{s \in I | \tau(s) \neq 0\}$ is empty, which leads a contradiction. This yields that $\alpha(s)$ is a plane curve, and hence, M is an open part of a plane. This completes the proof of Theorem 6. □

Note that a plane is a kind of cylindrical surface and also a kind of circular right cone. Thus, by summarizing all the results in this section, we established the following classification theorem for developable surfaces with generalized 1-type Gauss maps:

Theorem 4. (Classification Theorem) *A developable surface, M, in $\mathbb{E}^3$ has a generalized 1-type Gauss map if and only if it is an open part of one of the following:*

(1) A cylindrical surface,
(2) A circular right cone,
(3) A conical surface parameterized by

$$x(s,t) = \alpha_0 + t\beta(s),$$

where α_0 is a constant vector and $\beta(s)$ is a unit speed spherical curve in the unit sphere, $\mathbb{S}^2(1)$, with a positive geodesic curvature, κ_g, which is, for some indefinite integral, $F(t)$, of the function $\psi(t) = \left(a \ln t + b - t^2 - (\ln t)^2\right)^{-1/2}$ with $a, b \in R$, given by

$$\kappa_g(s) = \frac{1}{F^{-1}(\pm s)}.$$

4. Cylindrical Hypersurfaces with Generalized 1-Type Gauss Maps

In this section, we study cylindrical hypersurfaces with generalized 1-type Gauss maps in $\mathbb{E}^{n+2}$. Suppose that a hypersurface, M, in $\mathbb{E}^{n+2}$ has a generalized 1-type Gauss map, that is, the Gauss map, G, of the hypersurface satisfies the condition

$$\Delta G = fG + gC \tag{65}$$

for some non-zero smooth functions, f, g, and a non-zero constant vector, C. By combining (10) and (65) and taking the scalar product with the orthonormal local frame, $e_1, e_2, \ldots, e_{n+1}$ of M and the Gauss map, G, respectively, we obtain

$$(n+1)e_i(H) = gC_i, \quad i = 1, 2, \ldots, n+1 \tag{66}$$

and

$$||A_G||^2 = f + gC_{n+2}, \tag{67}$$

where, for $i = 1, 2, \ldots, n+1$, $C_i = \langle C, e_i \rangle$ and $C_{n+2} = \langle C, G \rangle$.

By extending Theorem 3.3, finally, we prove the following theorem:

Theorem 5. *A cylindrical hypersurface, M, in $\mathbb{E}^{n+2}$ has a generalized 1-type Gauss map.*

Proof. Let M be a cylindrical hypersurface in the $(n+2)$-dimensional Euclidean space, $\mathbb{E}^{n+2}$. Then, M can be parametrized by

$$x(s, t_1, \ldots, t_n) = \alpha(s) + \sum_{i=1}^{n} t_i \beta_i$$

such that $\langle \alpha', \alpha' \rangle = 1$, $\langle \alpha', \beta_i \rangle = 0$ and $\langle \beta_i, \beta_j \rangle = \delta_{ij}$, $i, j = 1, \ldots, n$. Then, the generator α is a plane curve with the Frenet frame T, N and we have the orthonormal frame $\{e_1, e_2, \ldots, e_{n+1}\}$ on M, such that $e_i = \frac{\partial}{\partial t_i}$, $i = 1, \ldots, n$ and $e_{n+1} = \frac{\partial}{\partial s} = T$. Hence, by rearranging β_i, if necessary, we may assume that the Gauss map, G, of M is given by $G = e_1 \times \cdots \times e_{n+1} = N$. By direct calculation, we get

$$\begin{aligned} &\widetilde{\nabla}_{e_i} e_j = \widetilde{\nabla}_{e_{n+1}} e_j = \widetilde{\nabla}_{e_i} e_{n+1} = 0,\ i, j = 1, \ldots n, \\ &\widetilde{\nabla}_{e_i} G = 0,\ i = 1, \ldots n, \quad \widetilde{\nabla}_{e_{n+1}} e_{n+1} = \kappa G, \quad \widetilde{\nabla}_{e_{n+1}} G = -\kappa e_{n+1}, \end{aligned} \tag{68}$$

where κ is the curvature function of the generator, α. (68) implies that

$$H = \frac{\kappa}{n+1}, \quad ||A_G||^2 = \kappa^2, \tag{69}$$

which are the functions of only s.

Now, suppose that M has a generalized 1-type Gauss map. That is, G satisfies (65). Then, C in $\mathbb{E}^{n+2}$ can be expressed as $C = \sum_{j=1}^{n+1} C_j e_j + C_{n+2} G$ in the frame $\{e_1, e_2, \ldots, e_{n+1}, G\}$. Together with (69), (66) implies that $C_i = 0$ because $e_i(H) = 0$, but $g \neq 0$ for $i = 1, \ldots, n$. Hence, we have

$$C = C_{n+1} e_{n+1} + C_{n+2} G = C_{n+1} T + C_{n+2} N. \tag{70}$$

By differentiating (70) with respect to e_i for $i = 1, \ldots, n$, (68) shows that

$$e_i(C_{n+1}) = e_i(C_{n+2}) = 0, \quad i = 1, \ldots, n. \tag{71}$$

Hence, C_{n+1} and C_{n+2} are functions of s only. By differentiating (70) with respect to e_{n+1}, (68) also gives

$$\begin{aligned} e_{n+1}\left(C_{n+1}\right) - \kappa(s) C_{n+2} &= 0, \\ e_{n+1}\left(C_{n+2}\right) + \kappa(s) C_{n+1} &= 0 \end{aligned} \tag{72}$$

with $C_{n+1}^2(s) + C_{n+2}^2(s) = d^2$ for some non-zero constant, d. Hence, we may put

$$C_{n+1}(s) = d \sin \theta(s), \quad C_{n+2}(s) = d \cos \theta(s), \tag{73}$$

where $\theta(s)$ is an indefinite integral of the curvature function $\kappa(s)$. Therefore, the constant vector, C, is given by

$$C = d \sin \theta(s) e_{n+1} + d \cos \theta(s) G = d \sin \theta(s) T + d \cos \theta(s) N. \tag{74}$$

Furthermore, it follows from (66), (67) and (69) that

$$\begin{aligned} f &= \kappa^2(s) - \kappa'(s) \cot \theta(s), \\ g &= \frac{\kappa'}{d \sin \theta(s)}. \end{aligned} \tag{75}$$

Conversely, for a cylindrical hypersurface, M, in $\mathbb{E}^{n+2}$, we may choose a curve, $\alpha(s)$, and n unit vectors $\beta_1, \ldots, \beta_n$ such that M is parametrized by

$$x\left(s, t_1, \ldots, t_n\right) = \alpha(s) + \sum_{i=1}^{n} t_i \beta_i$$

such that $\langle \alpha', \alpha' \rangle = 1$, $\langle \alpha', \beta_i \rangle = 0$ and $\langle \beta_i, \beta_j \rangle = \delta_{ij}$, $i, j = 1, \ldots, n$. For a non-zero constant, d, and an indefinite integral, $\theta(s)$, of the curvature function $\kappa(s)$ of α, we put

$$C = d \sin \theta(s) e_{n+1} + d \cos \theta(s) G, \tag{76}$$

where $e_i = \frac{\partial}{\partial t_i}$, $e_{n+1} = \frac{\partial}{\partial s}$ and $G = e_1 \times e_2 \times \cdots \times e_{n+1}$ for $i = 1, \ldots, n$. It follows from (68) that $\widetilde{\nabla}_{e_1} C = 0$ and $\widetilde{\nabla}_{e_2} C = 0$, and hence, C is a constant vector. Furthermore, it is straightforward to show that the Gauss map of M satisfies

$$\Delta G = fG + gC,$$

where f, g and C are given in (75) and (76), respectively. This shows that the cylindrical hypersurface has a generalized 1-type Gauss map. □

5. Conclusions

To find the best possible estimate of the total mean curvature of a compact submanifold of Euclidean space, Chen introduced the study of finite type submanifolds. Specifically, minimal submanifolds are characterized in a natural way. In our example, a cylindrical surface has neither a usual 1-type, nor a pointwise 1-type Gauss map. In this reason, we defined a new definiton, the generalized 1-type Gauss map. After that, we characterized developable surfaces with a generalized 1-type Gauss map in $\mathbb{E}^3$.

Author Contributions: D.W.Y. and J.W.L. gave the idea of establishing generalized finite type surfaces on Euclidean space. D.-S.K. and Y.H.K. checked and polished the draft.

Funding: The first author was supported by Basic Science Research Program through the National Research Foundation of Korea (NRF) funded by the Ministry of Education (2015R1D1A1A01060046). The second author was supported by Basic Science Research Program through the National Research Foundation of Korea (NRF) funded by the Ministry of Education (2015020387). The fourth author, the corresponding author, was supported by Basic Science Research Program through the National Research Foundation of Korea (NRF) funded by the Ministry of Education (2017R1D1A1B03033978).

Acknowledgments: We would like to thank the referee for the careful review and the valuable comments which really improved the paper.

Conflicts of Interest: The authors declare no conflict of interest.

References

1. Chen, B.-Y. On submanifolds of finite type. *Soochow J. Math.* **1983**, *9*, 65–81.
2. Chen, B.-Y. *Total Mean Curvature and Submanifolds of Finite Type*; World Scientific Publishing Company: Singapore, 1984.
3. Baikoussis, C. Ruled submanifolds with finite type Gauss map. *J. Geom.* **1994**, *49*, 42–45. [CrossRef]
4. Baikoussis, C.; Defever, F.; Koufogiorgos, T.; Verstraelen, L. Finite type immersions of flat tori into Euclidean spaces. *Proc. Edinb. Math. Soc.* **1995**, *38*, 413–420. [CrossRef]
5. Chen, B.-Y.; Piccinni, P. Submanifolds with finite type Gauss map. *Bull. Aust. Math. Soc.* **1987**, *35*, 161–186. [CrossRef]
6. Ki, U.-H.; Kim, D.-S.; Kim, Y.H.; Roh, Y.-M. Surfaces of revolution with pointwise 1-type Gauss map in Minkowski 3-space. *Taiwan. J. Math.* **2009**, *13*, 317–338. [CrossRef]
7. Dillen, F.; Pas, J.; Vertraelen, L. On surfaces of finite type in Euclidean 3-space. *Kodai Math. J.* **1990**, *13*, 10–21. [CrossRef]
8. Kim, D.-S.; Kim, Y.H. Some classification results on finite type ruled submanifolds in a Lorentz-Minkowski space. *Taiwan. J. Math.* **2012**, *16*, 1475–1488. [CrossRef]
9. Kim, D.-S.; Kim, Y.H.; Jung, S.M. Some classifications of ruled submanifolds in Minkowski space and their Gauss map. *Taiwan. J. Math.* **2014**, *18*, 1021–1040. [CrossRef]
10. Kim, D.-S.; Kim, Y.H.; Yoon, D.W. Characterization of generalized B-scrolls and cylinders over finite type curves. *Indian J. Pure Appl. Math.* **2003**, *34*, 1523–1532.
11. Kim, D.-S.; Kim, Y.H.; Yoon, D.W. Finite type ruled surfaces in Lorentz-Minkowski space. *Taiwan. J. Math.* **2007**, *11*, 1–13. [CrossRef]
12. Yoon, D.W. Rotation surfaces with finite type Gauss map in $\mathbb{E}^4$. *Indian J. Pure Appl. Math.* **2001**, *32*, 1803–1808.
13. Chen, B.-Y.; Morvan, J.M.; Nore, T. Energie, tension et ordre des applications a valeurs dans un espace euclide n. *CRAS Paris* **1985**, *301*, 123–126.

14. Chen, B.-Y.; Morvan, J.M.; Nore, T. Energie, tension and finite type maps. *Kodai Math. J.* **1986**, *9*, 408–418. [CrossRef]
15. Kim, Y.H.; Yoon, D.W. Ruled surfaces with pointwise 1-type Gauss map. *J. Geom. Phys.* **2000**, *34*, 191–205. [CrossRef]
16. Aksoyak, F.K.; Yayli, Y. Boost invariant surfaces with pointwise 1-type Gauss map in Minkowski 4-space $\mathbb{E}_1^4$. *Bull. Korean Math. Soc.* **2014**, *51*, 1863–1874. [CrossRef]
17. Arslan, K.; Bulca, B.; Milousheva, V. Meridian surfaces in $\mathbb{E}^4$ with pointwise 1-type Gauss map. *Bull. Korean Math. Soc.* **2014**, *51*, 911–922. [CrossRef]
18. Chen, B.-Y.; Choi, M.; Kim, Y.H. Surfaces of revolution with pointwise 1-type Gauss map. *J. Korean Math. Soc.* **2005**, *42*, 447–455. [CrossRef]
19. Choi, M.; Kim, D.-S.; Kim, Y.H.; Yoon, D.W. Circular cone and its Gauss map. *Colloq. Math.* **2012**, *129*, 203–210. [CrossRef]
20. Choi, M.; Kim, Y.H.; Yoon, D.W. Classification of ruled surfaces with pointwise 1-type Gauss map in Minkowski 3-space. *Taiwan. J. Math.* **2011**, *15*, 1141–1161. [CrossRef]
21. Dursun, U. Hypersurfaces with pointwise 1-type Gauss map. *Taiwan. J. Math.* **2007**, *11*, 1407–1416. [CrossRef]
22. Dursun, U. Flat surfaces in the Euclidean space $\mathbb{E}^3$ with pointwise 1-type Gauss map. *Bull. Malays. Math. Sci. Soc.* **2010**, *33*, 469–478.
23. Dursun, U.; Bektas, B. Spacelike rotational surfaces of elliptic, hyperbolic and parabolic types in Minkowski space $\mathbb{E}_1^4$ with pointwise 1-type Gauss map. *Math. Phys. Anal. Geom.* **2014**, *17*, 247–263. [CrossRef]
24. Dursun, U.; Turgay, N.C. General rotational surfaces in Euclidean space $\mathbb{E}^4$ with pointwise 1-type Gauss map. *Math. Commun.* **2012**, **17**, 71–81.
25. Kim, Y.H.; Yoon, D.W. On the Gauss map of ruled surfaces in Minkowski space. *Rocky Mt. J. Math.* **2005**, *35*, 1555–1581. [CrossRef]
26. Turgay, N.C. On the marginally trapped surfaces in 4-dimensional space-times with finite type Gauss map. *Gen. Relativ. Gravit.* **2014**, *46*, 1621. [CrossRef]
27. Yoon, D.W. Surfaces of revolution in the three dimensional pseudo-Galilean space. *Glas. Mat.* **2013**, *48*, 415–428. [CrossRef]
28. Vaisman, I. *A First Course in Differerential Geometry*; Dekker: New York, NY, USA, 1984.

mathematics

MDPI

Article

Inextensible Flows of Curves on Lightlike Surfaces

Zühal Küçükarslan Yüzbaşı [1] and Dae Won Yoon [2,*]

[1] Department of Mathematics, Fırat University, Elazig 23119, Turkey; zuhal2387@yahoo.com.tr
[2] Department of Mathematics Education and RINS, Gyeongsang National University, Jinju 52828, Korea
* Correspondence: dwyoon@gnu.ac.kr; Tel.: +82-55-772-2256

Received: 28 September 2018; Accepted: 26 October 2018; Published: 29 October 2018

Abstract: In this paper, we study inextensible flows of a curve on a lightlike surface in Minkowski three-space and give a necessary and sufficient condition for inextensible flows of the curve as a partial differential equation involving the curvatures of the curve on a lightlike surface. Finally, we classify lightlike ruled surfaces in Minkowski three-space and characterize an inextensible evolution of a lightlike curve on a lightlike tangent developable surface.

Keywords: inextensible flow; lightlike surface; ruled surface; Darboux frame

1. Introduction

It is well known that many nonlinear phenomena in physics, chemistry and biology are described by dynamics of shapes, such as curves and surfaces, and the time evolution of a curve and a surface has significance in computer vision and image processing. The time evolution of a curve and a surface is described by flows, in particular inextensible flows of a curve and a surface. Physically, inextensible flows give rise to motion, for which no strain energy is induced. The swinging motion of a cord of fixed length or of a piece of paper carried by the wind can be described by inextensible flows of a curve and a surface. Furthermore, the flows arise in the context of many problems in computer vision and computer animation [1–4].

Chirikjian and Burdick [1] studied applications of inextensible flows of a curve. In [5], the authors derived the time evolution equations for an inextensible flow of a space curve and also studied inextensible flows of a developable ruled surface. In [6], the author investigated the general description of the binormal motion of a spacelike and a timelike curve in a three-dimensional de Sitter space and gave some explicit examples of a binormal motion of the curves. Schief and Rogers [4] studied the binormal motions of curves with constant curvature and torsion. Many authors have studied geometric flow problems [7–11].

The outline of the paper is organized as follows: In Section 2, we give some geometric concepts in Minkowski space and present the pseudo-Darboux frames of a spacelike curve and a lightlike curve on a lightlike surface. In Sections 3 and 4, we study inextensible flows of a spacelike curve and a lightlike curve on a lightlike surface. In the last section, we classify lightlike ruled surfaces and study inextensible flows of lightlike tangent developable surfaces.

2. Preliminaries

The Minkowski three-space $\mathbb{R}^3_1$ is a real space $\mathbb{R}^3$ with the indefinite inner product $\langle \cdot\,,\cdot\rangle$ defined on each tangent space by:

$$\langle \mathbf{x}, \mathbf{y}\rangle = -x_0y_0 + x_1y_1 + x_2y_2,$$

where $\mathbf{x} = (x_0, x_1, x_2)$ and $\mathbf{y} = (y_0, y_1, y_2)$ are vectors in $\mathbb{R}^3_1$.

A nonzero vector $\mathbf{x}$ in $\mathbb{R}^3_1$ is said to be spacelike, timelike or lightlike if $\langle \mathbf{x},\mathbf{x}\rangle > 0$, $\langle \mathbf{x},\mathbf{x}\rangle < 0$ or $\langle \mathbf{x},\mathbf{x}\rangle = 0$, respectively. Similarly, an arbitrary curve $\gamma = \gamma(s)$ is spacelike, timelike or lightlike if all of its tangent vectors $\gamma'(s)$ are spacelike, timelike or lightlike, respectively. Here "prime" denotes the derivative with respect to the parameter s.

Let M be a lightlike surface in Minkowski three-space $\mathbb{R}^3_1$, that is the induced metric of M is degenerate. Then, a curve γ on M is spacelike or lightlike.

Case 1: If γ is a spacelike curve, we can reparametrize it by the arc length s. Therefore, we have the unit tangent vector $\mathbf{t}(s) = \gamma'(s)$ of $\gamma(s)$. Since M is a lightlike surface, we have a lightlike normal vector $\mathbf{n}$ along γ. Therefore, we can choose a vector $\mathbf{g}$ satisfying:

$$\langle \mathbf{n}, \mathbf{g} \rangle = 1, \quad \langle \mathbf{t}, \mathbf{g} \rangle = \langle \mathbf{g}, \mathbf{g} \rangle = 0.$$

Then, we have pseudo-orthonormal frames $\{\mathbf{t}, \mathbf{n}, \mathbf{g}\}$, which are called the Darboux frames along $\gamma(s)$. By standard arguments, we have the following Frenet formulae:

$$\frac{d}{ds}\begin{pmatrix} \mathbf{t}(s) \\ \mathbf{n}(s) \\ \mathbf{g}(s) \end{pmatrix} = \begin{pmatrix} 0 & \kappa_g(s) & \kappa_n(s) \\ -\kappa_n(s) & \tau_g(s) & 0 \\ -\kappa_g(s) & 0 & -\tau_g(s) \end{pmatrix} \begin{pmatrix} \mathbf{t}(s) \\ \mathbf{n}(s) \\ \mathbf{g}(s) \end{pmatrix}, \tag{1}$$

where $\kappa_n = \langle \mathbf{t}'(s), \mathbf{n}(s) \rangle$, $\kappa_g = \langle \mathbf{t}'(s), \mathbf{g}(s) \rangle$ and $\tau_g = -\langle \mathbf{n}(s), \mathbf{g}'(s) \rangle$.

Case 2: Let γ be a lightlike curve parametrized by a pseudo arc length parameter s on a lightlike surface M in $\mathbb{R}^3_1$. Since a normal vector $\mathbf{n}$ of a lightlike surface M is lightlike, we can choose a vector $\mathbf{g}$ such that:

$$\langle \mathbf{g}, \mathbf{g} \rangle = 1, \quad \langle \mathbf{t}, \mathbf{g} \rangle = \langle \mathbf{g}, \mathbf{n} \rangle = 0.$$

Furthermore, we consider:

$$\langle \mathbf{t}, \mathbf{n} \rangle = 1.$$

Then, we have pseudo-orthonormal Darboux frames $\{\mathbf{t}, \mathbf{n}, \mathbf{g}\}$ along a nongeodesic lightlike curve $\gamma(s)$ on M and get the following Frenet formulae:

$$\frac{d}{ds}\begin{pmatrix} \mathbf{t}(s) \\ \mathbf{n}(s) \\ \mathbf{g}(s) \end{pmatrix} = \begin{pmatrix} \kappa_n(s) & 0 & \kappa_g(s) \\ 0 & -\kappa_n(s) & \tau_g(s) \\ -\tau_g(s) & -\kappa_g(s) & 0 \end{pmatrix} \begin{pmatrix} \mathbf{t}(s) \\ \mathbf{n}(s) \\ \mathbf{g}(s) \end{pmatrix}, \tag{2}$$

where $\kappa_n = \langle \mathbf{t}'(s), \mathbf{n}(s) \rangle$, $\kappa_g = \langle \mathbf{t}'(s), \mathbf{g}(s) \rangle$ and $\tau_g = -\langle \mathbf{n}(s), \mathbf{g}'(s) \rangle$.

3. Inextensible Flows of a Spacelike Curve

We assume that $\gamma : [0, l] \times [0, w] \to M \subset \mathbb{R}^3_1$ is a one-parameter family of the smooth spacelike curve on a lightlike surface in $\mathbb{R}^3_1$, where l is the arc length of the initial curve. Let u be the curve parametrization variable, $0 \leq u \leq l$. We put $v = ||\frac{\partial \gamma}{\partial u}||$, from which the arc length of γ is defined by $s(u) = \int_0^u v du$. Furthermore, the operator $\frac{\partial}{\partial s}$ is given in terms of u by $\frac{\partial}{\partial s} = \frac{1}{v}\frac{\partial}{\partial u}$, and the arc length parameter is given by $ds = vdu$.

On the Darboux frames $\{\mathbf{t}, \mathbf{n}, \mathbf{g}\}$ of the spacelike curve γ on a lightlike surface M in $\mathbb{R}^3_1$, any flow of γ can be given by:

$$\frac{\partial \gamma}{\partial t} = f_1 \mathbf{t} + f_2 \mathbf{n} + f_3 \mathbf{g}, \tag{3}$$

where f_1, f_2, f_3 are scalar speeds of the spacelike curve γ on a lightlike surface M, respectively. We put $s(u, t) = \int_0^u v du$; it is called the arc length variation of γ. From this, the requirement that the curve is not subject to any elongation or compression can be expressed by the condition:

$$\frac{\partial}{\partial t} s(u, t) = \int_0^u \frac{\partial v}{\partial t} du = 0 \tag{4}$$

for all $u \in [0, l]$.

Definition 1. *A curve evolution $\gamma(u,t)$ and its flow $\frac{\partial \gamma}{\partial t}$ of a spacelike curve in $\mathbb{R}^3_1$ are said to be inextensible if:*

$$\frac{\partial}{\partial t}\left|\left|\frac{\partial \gamma}{\partial u}\right|\right| = 0.$$

Now, we give the arc length preserving condition for curve flows.

Theorem 1. *Let M be a lightlike surface in Minkowski three-space $\mathbb{R}^3_1$ and $\{\mathbf{t}, \mathbf{n}, \mathbf{g}\}$ be the Darboux frames of a spacelike curve γ on M. If $\frac{\partial \gamma}{\partial t} = f_1\mathbf{t} + f_2\mathbf{n} + f_3\mathbf{g}$ is a flow of γ on a lightlike surface M in $\mathbb{R}^3_1$, then we have the following equation:*

$$\frac{\partial v}{\partial t} = \frac{\partial f_1}{\partial u} - vf_2\kappa_n - vf_3\kappa_g. \tag{5}$$

Proof. From the definition of a spacelike curve γ, we have $v^2 = \left\langle \frac{\partial \gamma}{\partial u}, \frac{\partial \gamma}{\partial u} \right\rangle$. Since u and t are independent coordinates, $\frac{\partial}{\partial u}$ and $\frac{\partial}{\partial t}$ commute. Therefore, by differentiating v^2, we have:

$$\begin{aligned}
2v\frac{\partial v}{\partial t} &= \frac{\partial}{\partial t}\left\langle \frac{\partial \gamma}{\partial u}, \frac{\partial \gamma}{\partial u} \right\rangle \\
&= 2\left\langle \frac{\partial \gamma}{\partial u}, \frac{\partial}{\partial u}(\frac{\partial \gamma}{\partial t}) \right\rangle \\
&= 2\left\langle \frac{\partial \gamma}{\partial u}, \frac{\partial}{\partial u}(f_1\mathbf{t} + f_2\mathbf{n} + f_3\mathbf{g}) \right\rangle \\
&= 2v\left\langle \mathbf{t}, (\frac{\partial f_1}{\partial u} - vf_2\kappa_n - vf_3\kappa_g)\mathbf{t} + (\frac{\partial f_2}{\partial u} + vf_1\kappa_g + vf_2\tau_g)\mathbf{n} + (\frac{\partial f_3}{\partial u} + vf_1\kappa_n - vf_3\tau_g)\mathbf{g} \right\rangle \\
&= 2v\left(\frac{\partial f_1}{\partial u} - vf_2\kappa_n - vf_3\kappa_g\right).
\end{aligned}$$

This completes the proof. □

Corollary 1. *Let $\frac{\partial \gamma}{\partial t} = f_1\mathbf{t} + f_2\mathbf{n} + f_3\mathbf{g}$ be a flow of a spacelike curve γ on a lightlike surface M in $\mathbb{R}^3_1$. If the curve γ is a geodesic curve or an asymptotic curve, then the following equation holds, respectively:*

$$\frac{\partial v}{\partial t} = \frac{\partial f_1}{\partial u} - vf_2\kappa_n$$

or:

$$\frac{\partial v}{\partial t} = \frac{\partial f_1}{\partial u} - vf_3\kappa_g.$$

Theorem 2. (Necessary and sufficient condition for an inextensible flow)

Let $\frac{\partial \gamma}{\partial t} = f_1\mathbf{t} + f_2\mathbf{n} + f_3\mathbf{g}$ be a flow of a spacelike curve γ on a lightlike surface M in $\mathbb{R}^3_1$. Then, the flow is inextensible if and only if:

$$\frac{\partial f_1}{\partial s} = f_2\kappa_n + f_3\kappa_g. \tag{6}$$

Proof. Suppose that the flow of a spacelike curve γ on M is inextensible. From (4) and (5), we have:

$$\frac{\partial}{\partial t}s(u,t) = \int_0^u \frac{\partial v}{\partial t}du = \int_0^u \left(\frac{\partial f_1}{\partial u} - vf_2\kappa_n - vf_3\kappa_g\right) du = 0.$$

It follows that:

$$\frac{\partial f_1}{\partial u} = vf_2\kappa_n + vf_3\kappa_g.$$

Since $\frac{\partial}{\partial s} = \frac{1}{v}\frac{\partial}{\partial u}$, we can obtain (6).

Conversely, by following a similar way as above, the proof is completed. □

Theorem 3. *Let $\frac{\partial \gamma}{\partial t} = f_1\mathbf{t} + f_2\mathbf{n} + f_3\mathbf{g}$ be a flow of a spacelike curve γ on a lightlike surface M in $\mathbb{R}^3_1$. If the flow is inextensible, then a time evolution of the Darboux frame $\{\mathbf{t}, \mathbf{n}, \mathbf{g}\}$ along a curve γ on a lightlike surface M is given by:*

$$\frac{d}{dt}\begin{pmatrix} \mathbf{t} \\ \mathbf{n} \\ \mathbf{g} \end{pmatrix} = \begin{pmatrix} 0 & \varphi_1 & \varphi_2 \\ -\varphi_2 & \varphi_3 & 0 \\ -\varphi_1 & 0 & -\varphi_3 \end{pmatrix} \begin{pmatrix} \mathbf{t} \\ \mathbf{n} \\ \mathbf{g} \end{pmatrix}, \tag{7}$$

where:

$$\begin{aligned} \varphi_1 &= \frac{\partial f_2}{\partial s} + f_1\kappa_g + f_2\tau_g, \\ \varphi_2 &= \frac{\partial f_3}{\partial s} + f_1\kappa_n - f_3\tau_g, \\ \varphi_3 &= \langle \frac{\partial \mathbf{n}}{\partial t}, \mathbf{g} \rangle. \end{aligned} \tag{8}$$

Proof. Noting that:

$$\begin{aligned} \frac{\partial \mathbf{t}}{\partial t} &= \frac{\partial}{\partial t}\left(\frac{\partial \gamma}{\partial s}\right) = \frac{\partial}{\partial s}(f_1\mathbf{t} + f_2\mathbf{n} + f_3\mathbf{g}) \\ &= \left(\frac{\partial f_2}{\partial s} + f_1\kappa_g + f_2\tau_g\right)\mathbf{n} + \left(\frac{\partial f_3}{\partial s} + f_1\kappa_n - f_3\tau_g\right)\mathbf{g}. \end{aligned} \tag{9}$$

On the other hand,

$$\begin{aligned} 0 &= \frac{\partial}{\partial t}\langle \mathbf{t}, \mathbf{n} \rangle = \langle \frac{\partial \mathbf{t}}{\partial t}, \mathbf{n} \rangle + \langle \mathbf{t}, \frac{\partial \mathbf{n}}{\partial t} \rangle = \frac{\partial f_3}{\partial s} + f_1\kappa_n - f_3\tau_g + \langle \mathbf{t}, \frac{\partial \mathbf{n}}{\partial t} \rangle \\ 0 &= \frac{\partial}{\partial t}\langle \mathbf{t}, \mathbf{g} \rangle = \langle \frac{\partial \mathbf{t}}{\partial t}, \mathbf{g} \rangle + \langle \mathbf{t}, \frac{\partial \mathbf{g}}{\partial t} \rangle = \frac{\partial f_2}{\partial s} + f_1\kappa_g + f_2\tau_g + \langle \mathbf{t}, \frac{\partial \mathbf{n}}{\partial t} \rangle \end{aligned}$$

because of $\langle \mathbf{n}, \mathbf{n} \rangle = \langle \mathbf{g}, \mathbf{g} \rangle = 0$ and $\langle \mathbf{n}, \mathbf{g} \rangle = 1$.

Thus, we have:

$$\frac{\partial \mathbf{n}}{\partial t} = -\left(\frac{\partial f_3}{\partial s} + f_1\kappa_n - f_3\tau_g\right)\mathbf{t} + \varphi_3\mathbf{n}, \tag{10}$$

$$\frac{\partial \mathbf{g}}{\partial t} = -\left(\frac{\partial f_2}{\partial s} + f_1\kappa_g + f_2\tau_g\right)\mathbf{t} - \varphi_3\mathbf{g}, \tag{11}$$

where $\varphi_3 = \langle \frac{\partial \mathbf{n}}{\partial t}, \mathbf{g} \rangle$. This completes the proof. □

Now, by using Theorem 3, we give the time evolution equations of the geodesic curvature, the normal curvature and the geodesic torsion of a spacelike curve on a lightlike surface.

Theorem 4. *Let $\frac{\partial \gamma}{\partial t} = f_1\mathbf{t} + f_2\mathbf{n} + f_3\mathbf{g}$ be a flow of a spacelike curve γ on a lightlike surface M in $\mathbb{R}^3_1$. Then, the time evolution equations of the functions κ_g, κ_n and τ_g for the inextensible spacelike curve γ are given by:*

$$\begin{aligned} \frac{\partial \kappa_g}{\partial t} &= \frac{\partial \varphi_1}{\partial s} + \varphi_1\tau_g - \varphi_3\kappa_g, \\ \frac{\partial \kappa_n}{\partial t} &= \frac{\partial \varphi_2}{\partial s} - \varphi_2\tau_g + \varphi_3\kappa_n \\ \frac{\partial \tau_g}{\partial t} &= \frac{\partial \varphi_3}{\partial s} + \varphi_1\kappa_n - \varphi_2\kappa_g + 2\varphi_3\tau_g. \end{aligned} \tag{12}$$

Proof. It is well known that the arc length and time derivatives commute. This implies the inextensibility of γ. Accordingly, the compatibility conditions are $\frac{\partial}{\partial s}\left(\frac{\partial \mathbf{t}}{\partial t}\right) = \frac{\partial}{\partial t}\left(\frac{\partial \mathbf{t}}{\partial s}\right)$, etc. On the other hand,

$$\begin{aligned}\frac{\partial}{\partial s}\left(\frac{\partial \mathbf{t}}{\partial t}\right) &= \frac{\partial}{\partial s}(\varphi_1 \mathbf{n} + \varphi_2 \mathbf{g}) \\ &= (-\varphi_1 \kappa_n - \varphi_2 \kappa_g)\mathbf{t} + (\frac{\partial \varphi_1}{\partial s} + \varphi_1 \tau_g)\mathbf{n} + (\frac{\partial \varphi_2}{\partial s} - \varphi_2 \tau_g)\mathbf{g},\end{aligned}$$

and:

$$\begin{aligned}\frac{\partial}{\partial t}\left(\frac{\partial \mathbf{t}}{\partial s}\right) &= \frac{\partial}{\partial t}(\kappa_g \mathbf{n} + \kappa_n \mathbf{g}) \\ &= (-\varphi_1 \kappa_n - \varphi_2 \kappa_g)\mathbf{t} + (\frac{\partial \kappa_g}{\partial t} + \varphi_3 \kappa_g)\mathbf{n} + (\frac{\partial \kappa_n}{\partial t} - \varphi_3 \kappa_n)\mathbf{g}.\end{aligned}$$

Comparing the two equations, we find:

$$\begin{aligned}\frac{\partial \kappa_g}{\partial t} &= \frac{\partial \varphi_1}{\partial s} + \varphi_1 \tau_g - \varphi_3 \kappa_g, \\ \frac{\partial \kappa_n}{\partial t} &= \frac{\partial \varphi_2}{\partial s} - \varphi_2 \tau_g + \varphi_3 \kappa_n.\end{aligned}$$

It follows from (8) that we can obtain the first and the second equation of (12).

Furthermore by using $\frac{\partial}{\partial s}\left(\frac{\partial \mathbf{n}}{\partial t}\right) = \frac{\partial}{\partial t}\left(\frac{\partial \mathbf{n}}{\partial s}\right)$ and following a similar way as above, we can obtain the third equation of (12). The proof is completed. □

Remark 1. *As applications of inextensible flows of a spacelike curve on a lightlike surface, we can consider geometric phases of the repulsive-type nonlinear Schödinger equation (NLS^-) (cf. [12]).*

4. Inextensible Flows of a Lightlike Curve

Let γ be a lightlike curve on a lightlike surface M in $\mathbb{R}^3_1$. We note that a lightlike curve $\gamma(u)$ satisfies $\langle \gamma''(u), \gamma''(u) \rangle \geq 0$. We say that a lightlike curve $\gamma(u)$ is parametrized by the pseudo arc length if $\langle \gamma''(u), \gamma''(u) \rangle = 1$. If a lightlike curve $\gamma(u)$ satisfies $\langle \gamma''(u), \gamma''(u) \rangle \neq 0$, then $\langle \gamma''(u), \gamma''(u) \rangle > 0$, and:

$$s(u) = \int_0^u \langle \gamma''(u), \gamma''(u) \rangle^{\frac{1}{4}} du$$

becomes the pseudo arc length parameter. Let us consider a lightlike curve $\gamma(u)$ on a lightlike surface M in $\mathbb{R}^3_1$ with $\langle \gamma''(u), \gamma''(u) \rangle \neq 0$.

Let $\gamma : [0, l] \times [0, w] \to M \subset \mathbb{R}^3_1$ be a one-parameter family of smooth lightlike curves on a lightlike surface in $\mathbb{R}^3_1$, where l is the arc length of the initial curve. We put $v^4 = \langle \gamma''(u), \gamma''(u) \rangle$, from which the pseudo arc length of γ is defined by $s(u) = \int_0^u v du$. Furthermore, the operator $\frac{\partial}{\partial s}$ is given in terms of u by $\frac{\partial}{\partial s} = \frac{1}{v}\frac{\partial}{\partial u}$, and the pseudo arc length parameter is given by $ds = v du$.

On the other hand, a flow $\frac{\partial \gamma}{\partial t}$ of γ can be given by:

$$\frac{\partial \gamma}{\partial t} = f_1 \mathbf{t} + f_2 \mathbf{n} + f_3 \mathbf{g} \tag{13}$$

in terms of the Darboux frames $\{\mathbf{t}, \mathbf{n}, \mathbf{g}\}$ of the lightlike curve γ on a lightlike surface M in $\mathbb{R}^3_1$, where f_1, f_2, f_3 are scalar speeds of the lightlike curve γ, respectively. We put $s(u,t) = \int_0^u v du$, it is called the pseudo arc length variation of γ. From this, we have the following condition:

$$\frac{\partial}{\partial t} s(u,t) = \int_0^u \frac{\partial v}{\partial t} du = 0 \tag{14}$$

for all $u \in [0, l]$.

Definition 2. *A curve evolution $\gamma(u,t)$ and its flow $\frac{\partial\gamma}{\partial t}$ of a lightlike curve γ in $\mathbb{R}^3_1$ are said to be inextensible if:*

$$\frac{\partial}{\partial t}\left\langle \frac{\partial^2\gamma}{\partial u^2}, \frac{\partial^2\gamma}{\partial u^2} \right\rangle^{\frac{1}{4}} = 0.$$

Theorem 5. *Let M be a lightlike surface in Minkowski three-space $\mathbb{R}^3_1$ and $\{\mathbf{t},\mathbf{n},\mathbf{g}\}$ be the Darboux frames along a lightlike curve γ on M. If $\frac{\partial\gamma}{\partial t} = f_1\mathbf{t} + f_2\mathbf{n} + f_3\mathbf{g}$ is a flow of γ on a lightlike surface M, then we have the following equation:*

$$\frac{\partial v}{\partial t} = \frac{1}{2v^3}\left[(\frac{\partial v}{\partial u} + v^2\kappa_n)\left(\frac{\partial \Phi_2}{\partial u} - v\kappa_n\Phi_2 - v\kappa_g\Phi_3\right) + v^2\kappa_g\left(\frac{\partial\Phi_3}{\partial u} + v\kappa_g\Phi_1 + v\tau_g\Phi_2\right)\right], \quad (15)$$

where:

$$\Phi_1 = \frac{\partial f_1}{\partial u} + vf_1\kappa_n - vf_3\tau_g,$$
$$\Phi_2 = \frac{\partial f_2}{\partial u} - vf_2\kappa_n - vf_3\kappa_g,$$
$$\Phi_3 = \frac{\partial f_1}{\partial u} + vf_1\kappa_g + vf_2\tau_g.$$

Proof. From the definition of a lightlike curve γ, we have $v^4 = \left\langle \frac{\partial^2\gamma}{\partial u^2}, \frac{\partial^2\gamma}{\partial u^2}\right\rangle$. By differentiating v^4, we have:

$$4v^3\frac{\partial v}{\partial t} = \frac{\partial}{\partial t}\left\langle \frac{\partial^2\gamma}{\partial u^2}, \frac{\partial^2\gamma}{\partial u^2}\right\rangle = 2\left\langle \frac{\partial^2\gamma}{\partial u^2}, \frac{\partial^2}{\partial u^2}(\frac{\partial\gamma}{\partial t})\right\rangle. \quad (16)$$

On the other hand,

$$\frac{\partial^2\gamma}{\partial u^2} = \frac{\partial}{\partial u}\left(\frac{\partial\gamma}{\partial u}\right) = \frac{\partial}{\partial u}(v\mathbf{t}) = \left(\frac{\partial v}{\partial u} + v^2\kappa_n\right)\mathbf{t} + v^2\kappa_g\mathbf{g}$$

and:

$$\frac{\partial^2}{\partial u^2}(\frac{\partial\gamma}{\partial t}) = \frac{\partial^2}{\partial u^2}(f_1\mathbf{t} + f_2\mathbf{n} + f_3\mathbf{g})$$
$$= \left[\frac{\partial\Phi_1}{\partial u} + v\kappa_n\Phi_1 - v\tau_g\Phi_3\right]\mathbf{t} + \left[\frac{\partial\Phi_2}{\partial u} - v\kappa_n\Phi_2 - v\kappa_g\Phi_3\right]\mathbf{n} + \left[\frac{\partial\Phi_3}{\partial u} + v\kappa_g\Phi_1 + v\tau_g\Phi_2\right]\mathbf{g}.$$

Thus, (16) implies (15). This completes the proof. □

Theorem 6. *Let $\frac{\partial\gamma}{\partial t} = f_1\mathbf{t} + f_2\mathbf{n} + f_3\mathbf{g}$ be a flow of a lightlike curve γ on a lightlike surface M in $\mathbb{R}^3_1$. Then, the flow is inextensible if and only if:*

$$\left(\frac{\partial v}{\partial s} + v\kappa_n\right)\frac{\partial\Phi_2}{\partial s} + v\kappa_g\frac{\partial\Phi_3}{\partial s} = \left(\frac{\partial v}{\partial s} + v\kappa_n\right)(\kappa_n\Phi_2 + \kappa_g\Phi_3) - v\kappa_g(\kappa_g\Phi_1 + \tau_g\Phi_2). \quad (17)$$

Proof. Suppose that the flow of a lightlike curve γ on M is inextensible. By using (15) and $\frac{\partial}{\partial s} = \frac{1}{v}\frac{\partial}{\partial u}$, (14) gives (17). Conversely, by following a similar way as above, the proof is completed. □

Next, we give the time evolution equations of the Darboux frame of a lightlike curve on a lightlike surface.

Theorem 7. *Let $\frac{\partial\gamma}{\partial t} = f_1\mathbf{t} + f_2\mathbf{n} + f_3\mathbf{g}$ be a flow of a lightlike curve γ on a lightlike surface M in $\mathbb{R}^3_1$. If the flow is inextensible, then a time evolution of the Darboux frame $\{\mathbf{t}, \mathbf{n}, \mathbf{g}\}$ along a curve γ on a lightlike surface M is given by:*

$$\frac{d}{dt}\begin{pmatrix}\mathbf{t}\\ \mathbf{n}\\ \mathbf{g}\end{pmatrix} = \begin{pmatrix}\frac{\Phi_1}{v} & 0 & \frac{\Phi_3}{v}\\ 0 & -\frac{\Phi_1}{v} & \Theta\\ -\Theta & -\frac{\Phi_3}{v} & 0\end{pmatrix}\begin{pmatrix}\mathbf{t}\\ \mathbf{n}\\ \mathbf{g}\end{pmatrix}, \tag{18}$$

where $\Theta = \langle\frac{\partial \mathbf{n}}{\partial t}, \mathbf{g}\rangle$.

Proof. The proof can be obtained by using a similar method of proof of Theorem 3. □

Theorem 8. *Let $\frac{\partial\gamma}{\partial t} = f_1\mathbf{t} + f_2\mathbf{n} + f_3\mathbf{g}$ be a flow of a lightlike curve γ on a lightlike surface M in $\mathbb{R}^3_1$. Then, the time evolution equations of the functions κ_g, κ_n and τ_g for the inextensible spacelike curve γ are given by:*

$$\begin{aligned}\frac{\partial\kappa_g}{\partial t} &= \frac{\partial}{\partial s}(\frac{1}{v}\Phi_3) + \frac{1}{v}\left(\kappa_g\Phi_1 - \kappa_n\Phi_3\right),\\ \frac{\partial\kappa_n}{\partial t} &= \frac{\partial}{\partial s}(\frac{1}{v}\Phi_1) + \kappa_g\Theta - \frac{1}{v}\tau_g\Phi_3,\\ \frac{\partial\tau_g}{\partial t} &= \frac{\partial\Theta}{\partial s} + \kappa_n\Theta - \frac{1}{v}\tau_g\Phi_1.\end{aligned} \tag{19}$$

Proof. The proof can be obtained by using a similar method of proof of Theorem 4. □

5. Lightlike Ruled Surfaces

In this section, we investigate inextensible flows of ruled surfaces, in particular lightlike ruled surfaces in Minkowski three-space $\mathbb{R}^3_1$.

Let I be an open interval on the real line $\mathbb{R}$. Let α be a curve in $\mathbb{R}^3_1$ defined on I and β a transversal vector field along α. For an open interval J of $\mathbb{R}$, we have the parametrization for M:

$$X(u,v) = \alpha(u) + v\beta(u), \quad u \in I, \quad v \in J.$$

Here, α is called a base curve and β a director vector field. In particular, the director vector field β can be naturally chosen so that it is orthogonal to α, that is $\langle\alpha', \beta\rangle = 0$. It is well known that the ruled surface is developable if $\det(\alpha'\beta\beta')$ is identically zero. A developable surface is a surface whose Gaussian curvature of the surface is everywhere zero.

On the other hand, the tangent vectors are given by:

$$X_u = \frac{\partial X}{\partial u} = \alpha'(u) + v\beta'(u), \quad X_v = \frac{\partial X}{\partial v} = \beta(u),$$

which imply that the coefficients of the first fundamental form of the surface are given by:

$$\begin{aligned}E &= \langle X_u, X_u\rangle = \langle\alpha', \alpha'\rangle + 2v\langle\alpha', \beta'\rangle + v^2\langle\beta', \beta'\rangle,\\ F &= \langle X_u, X_v\rangle = 0,\\ G &= \langle X_v, X_v\rangle = \langle\beta, \beta\rangle.\end{aligned}$$

Suppose that the ruled surface is lightlike. Then, we get $E = 0$ or $G = 0$.

First of all, we consider $E = 0$; it implies that:

$$\langle\alpha', \alpha'\rangle = 0, \quad \langle\alpha', \beta'\rangle = 0, \quad \langle\beta', \beta'\rangle = 0. \tag{20}$$

Thus, a base curve α is lightlike, and a director vector β is constant or β' is lightlike.

Case 1: If β is constant, from $\langle \alpha', \beta \rangle = 0$, β is a lightlike vector or a spacelike vector. If β is lightlike, there exists a smooth function k such that $\beta = k\alpha'$. This is a contradiction because $G = 0$. If β is spacelike as a constant vector, then the lightlike cylindrical ruled surface is parametrized by:

$$X(u,v) = \alpha(u) + v\beta,$$

where α is a lightlike curve and β is a constant spacelike vector.

Case 2: Let β' be a lightlike vector. Since $\langle \alpha', \beta' \rangle = 0$, there exists a smooth function k such that $\beta' = k\alpha'$. Thus, a lightlike non-cylindrical ruled surface is parametrized by:

$$X(u,v) = \alpha(u) + v\beta(u), \tag{21}$$

where α and β satisfy the condition (20).

Next, we consider $G = \langle \beta, \beta \rangle = 0$, since $\beta \neq \mathbf{0}$, a director vector β must be lightlike. Furthermore, since $\langle \alpha', \beta \rangle = 0$, α is a spacelike curve or a lightlike curve.

Case 1: If α is a spacelike curve, then a lightlike non-cylindrical ruled surface is parametrized by:

$$X(u,v) = \alpha(u) + v\beta(u), \tag{22}$$

where α is a spacelike curve and β is a lightlike vector.

Case 2: Let α be a lightlike curve. Then, there exists a smooth function k such that $\beta' = k\alpha'$, and a lightlike ruled surface as a tangent developable surface is parametrized by:

$$X(u,v) = \alpha(u) + vk\alpha'(u), \tag{23}$$

where α and α'' are a lightlike curve and a spacelike vector, respectively.

In [5], the authors gave the following:

Definition 3. *A surface evolution $X(u,v,t)$ and its flow $\frac{\partial X}{\partial t}$ are said to be inextensible if the coefficients of the first fundamental form of the surface satisfy:*

$$\frac{\partial E}{\partial t} = \frac{\partial F}{\partial t} = \frac{\partial G}{\partial t} = 0.$$

This definition states that the surface $X(u,v,t)$ is, for all time t, the isometric image of the original surface $X(u,v,t_0)$ defined at some initial time t_0.

Now, we study inextensible flows of a lightlike tangent developable surface in Minkowski three-space.

Consider a lightlike tangent developable surface parametrized by:

$$X(u,v) = \alpha(u) + v\alpha'(u), \tag{24}$$

where α is a lightlike curve. Suppose that the parameter u is a pseudo-arc length of α. In this case, we get $E = v^2||\alpha''||^2$ and $F = G = 0$.

Thus, we have:

Theorem 9. *Let $X(u,v)$ be a lightlike tangent developable surface given by (24). The surface evolution $X(u,v,t) = \alpha(u,t) + v\alpha'(u,t)$ is inextensible if and only if:*

$$\frac{\partial}{\partial t}||\alpha''||^2 = 0.$$

As a consequence, we have the following results:

Theorem 10. *Let $X(u,v,t) = \alpha(u,t) + v\alpha'(u,t)$ be a surface evolution of a lightlike tangent developable surface given by (24) in $\mathbb{R}^3_1$. Then, we have the following statements:*

(1) *$\alpha(u,t)$ is an inextensible evolution of a lightlike curve $\alpha(u)$ in $\mathbb{R}^3_1$.*

(2) *An inextensible evolution of a lightlike tangent developable surface can be completely characterized by the inextensible evolutions of a lightlike curve $\alpha(u)$ in $\mathbb{R}^3_1$.*

Proof. In fact, $0 = \frac{\partial}{\partial t}||\alpha''||^2 = 2||\alpha''||\frac{\partial}{\partial t}||\alpha''||$ and $\alpha'' \neq \mathbf{0}$, and we get $\frac{\partial}{\partial t}||\alpha''|| = 0$; it implies $\frac{\partial}{\partial t}||\alpha''||^{\frac{1}{2}} = 0$. This means that $\alpha(u,t)$ satisfies the condition for Definition 2. □

Theorem 11. *Let $X(u,v,t) = \alpha(u,t) + v\alpha'(u,t)$ be a surface evolution of a lightlike tangent developable surface given by (24) in $\mathbb{R}^3_1$, and $\frac{\partial \alpha}{\partial t} = f_1\mathbf{t} + f_2\mathbf{n} + f_3\mathbf{g}$, where $\mathbf{t}, \mathbf{n}, \mathbf{g}$ are the Darboux frames along a lightlike curve α on a lightlike surface. If the surface evolution $X(u,v,t)$ is inextensible, then f_1, f_2, f_3 satisfy Equation (19).*

6. Conclusions

We study an inextensible flow of a spacelike or a lightlike curve on a lightlike surface in Minkowski three-space and investigate a time evolution of the Darboux frame $\{\mathbf{t}, \mathbf{n}, \mathbf{g}\}$ (see Theorems 3 and 7) and the functions κ_n, κ_g and τ_g (see Theorems 4 and 8). Furthermore, in Theorems 2 and 6, we give a necessary and sufficient condition of inextensible flows of a spacelike curve and a lightlike curve on a lightlike surface in terms of a partial differential equation involving the curvatures of the curve on a lightlike surface. Finally, we completely classify lightlike ruled surfaces in Minkowski three-space and characterize an inextensible evolution of a lightlike curve on a lightlike tangent developable surface (see Theorems 9 and 10).

Author Contributions: D.W.Y. gave the idea of inextensible flows of a spacelike curve and a lightlike curve on a lightlike surface. Z.K.Y. checked and polished the draft.

Funding: The second author was supported by the Basic Science Research Program through the National Research Foundation of Korea (NRF) funded by the Ministry of Education (NRF-2018R1D1A1B07046979).

Conflicts of Interest: The authors declare no conflict of interest.

References

1. Chirikjian, G.S.; Burdick, J.W. Kinematics of hyper-redundant manipulation. In Proceedings of the ASME Mechanisms Conference, Chicago, IL, USA, 16–19 September 1990; pp. 391–396.
2. Desbrun, M.; Cani-Gascuel, M.-P. Active implicit surface for animation. In *Graphics Interface*; The Canadian Information Processing Society: Mississauga, ON, Canada, 1998; pp. 143–150.
3. Kass, M.; Witkin, A.; Terzopoulos, D. Snakes: Active contour models. In Proceedings of the 1st International Conference on Computer Vision, London, UK, 8–11 June 1987; pp. 259–268.
4. Schief, W.K.; Rogers, C. Binormal motion of curves of constant curvature and torsion. generation of soliton surfaces. *Proc. R. Soc. Lond. A* **1999**, *455*, 3163–3188. [CrossRef]
5. Kwon, D.Y.; Park, F.C. Inextensible flows of curves and developable surfaces. *Appl. Math. Lett.* **2005**, *18*, 1156–1162. [CrossRef]
6. Mohamed, S.G. Binormal motions of inextensible curves in de-sitter space $\mathbb{S}^{2,1}$. *J. Egypt. Math. Soc.* **2017**, *25*, 313–318. [CrossRef]
7. Andrews, B. Classification of limiting shapes for isotropic curve flows. *J. Am. Math. Soc.* **2003**, *16*, 443–459. [CrossRef]
8. Gurbuz, N. Inextensible flows of spacelike, timelike and null curves. *Int. J. Contemp. Math. Sci.* **2009**, *4*, 1599–1604.
9. Hussien, R.A.; Mohamed, S.G. Generated surfaces via inextensible flows of curves in $\mathbb{R}^3$. *J. Appl. Math.* **2016**, *2016*, 6178961. [CrossRef]
10. Yeneroglu, M. On new characterization of inextensible flows of space-like curves in de Sitter space. *Open Math.* **2016**, *14*, 946–954. [CrossRef]

11. Zhu, X.-P. Asymptotic behavior of anisotropic curve flows. *J. Differ. Geom.* **1998**, *48*, 225–274. [CrossRef]
12. Gurbuz, N. Three clasess of non-lightlike curve evolution according to Darboux frame and geometric phase. *Int. J. Geom. Methods Mod. Phys.* **2018**, *15*, 1850023. [CrossRef]

mathematics

MDPI

Article

Trans-Sasakian 3-Manifolds with Reeb Flow Invariant Ricci Operator

Yan Zhao [1], Wenjie Wang [2,*] and Ximin Liu [2]

1. Department of Mathematics, College of Science, Henan University of Technology, Zhengzhou 450001, Henan, China; zy4012006@126.com
2. Wenjie Wang, School of Mathematical Sciences, Dalian University of Technology, Dalian 116024, Liaoning, China; ximinliu@dlut.edu.cn

* Correspondence: wangwj072@163.com

Received: 19 September 2018; Accepted: 6 November 2018; Published: 9 November 2018

Abstract: Let M be a three-dimensional trans-Sasakian manifold of type (α, β). In this paper, we obtain that the Ricci operator of M is invariant along Reeb flow if and only if M is an α-Sasakian manifold, cosymplectic manifold or a space of constant sectional curvature. Applying this, we give a new characterization of proper trans-Sasakian 3-manifolds.

Keywords: trans-Sasakian 3-manifold; Reeb flow symmetry; Ricci operator

---

1. Introduction

A trans-Sasakian manifold is usually denoted by $(M, \phi, \xi, \eta, g, \alpha, \beta)$, where both α and β are smooth functions and (ϕ, ξ, η, g) is an almost contact metric structure. M is said to be proper if either $\alpha = 0$ or $\beta = 0$. When $\beta = 0$, α is a constant if $\dim M \geq 5$ (see [1]) and in this case M becomes an α-Sasakian manifold if $\alpha \in \mathbb{R}^*$ or a cosymplectic manifold if $\alpha = 0$. This conclusion is not necessarily true for dimension three. However, unlike the above case, when $\alpha = 0$, β is not necessarily a constant even if $\dim M \geq 5$ or M is compact for dimension three (see [2]). The set of all trans-Sasakian manifolds of type $(0, \beta)$ coincides with that of all f-cosymplectic manifolds (see [3]) or f-Kenmotsu manifolds (see [4–6]). A trans-Sasakian manifold of dimension ≥ 5 must be proper (see [1]). In the geometry of trans-Sasakian 3-manifolds, there exists a basic interesting problem, that is:

Under what condition is a trans-Sasakian 3-manifold proper?

De [7–12], Deshmukh [13–15], Wang and Liu [16] and Wang [2,17] answered this question from various points of view. In this paper, we study this question under a new geometric condition. Before stating our main results, we recall some results related with such a condition.

On an almost contact metric manifold (M, ϕ, ξ, η, g), the Ricci operator of M is said to be Reeb flow invariant if it satisfies

$$\mathcal{L}_\xi Q = 0, \tag{1}$$

where $\mathcal{L}$, ξ and Q are the Lie derivative, Reeb vector field and the Ricci operator, respectively. Cho in [18] proved that a contact metric 3-manifold satisfies Equation (1) if and only if it is Sasakian or locally isometric to $SU(2)$ (or $SO(3)$), $SL(2,R)$ (or $O(1,2)$), the group $E(2)$ of rigid motions of Euclidean 2-plane. Cho in [19] proved that an almost cosymplectic 3-manifold satisfies (1) if and only if it is either cosymplectic or locally isometric to the group $E(1,1)$ of rigid motions of Minkowski 2-space. In addition, Cho and Kimura in [20] proved that an almost Kenmotsu 3-manifold satisfies (1) if and only if it is of constant sectional curvature -1 or a non-unimodular Lie group. Reeb flow invariant Ricci operators were also investigated on the unit tangent sphere bundle of a Riemannian manifold (see [21]), even on real hypersurfaces in complex two-plane Grassmannians (see [22]). In this paper, we obtain a new characterization of proper trans-Sasakian 3-manifolds by employing (1) and proving

Theorem 1. *The Ricci operator of a trans-Sasakian 3-manifold is invariant along Reeb flow if and only if the manifold is an α-Sasakian manifold, cosymplectic manifold or a space of constant sectional curvature.*

According to calculations shown in Section 3, we observe that Ricci parallelism with respect to the Levi–Civita connection (i.e., $\nabla Q = 0$) is stronger than a Reeb flow invariant Ricci operator. Thus, we have

Remark 1. *Theorem 1 is an extension of Wang and Liu [16] (Theorem 3.12).*

Some corollaries induced from Theorem 1 are also given in the last section.

2. Trans-Sasakian Manifolds

On a smooth Riemannian manifold (M, g) of dimension $2n+1$, we assume that ϕ, ξ and η are $(1,1)$-type, $(1,0)$-type and $(0,1)$-type tensor fields, respectively. According to [23], M is called an almost contact metric manifold if

$$\begin{aligned} \phi^2 X = -X + \eta(X)\xi,\ \eta(\xi) = 1,\ \eta(\phi X) = 0, \\ g(\phi X, \phi Y) = g(X, Y) - \eta(X)\eta(Y),\ \eta(X) = g(X, \xi) \end{aligned} \tag{2}$$

for any vector fields X and Y. An almost contact metric manifold is said to be normal if $[\phi, \phi] = -2\mathrm{d}\eta \otimes \xi$, where $[\phi, \phi]$ denotes the Nijenhuis tensor of ϕ.

A normal almost contact metric manifold is called a *trans-Sasakian manifold* (see [1]) if

$$(\nabla_X \phi)Y = \alpha(g(X, Y)\xi - \eta(Y)X) + \beta(g(\phi X, Y)\xi - \eta(Y)\phi X) \tag{3}$$

for any vector fields X, Y and two smooth functions α, β. In particular, a three-dimensional almost contact metric manifold is trans-Sasakian if and only if it is normal (see [24,25]).

A normal almost contact metric manifold is called an *α-Sasakian manifold* if $\mathrm{d}\eta = \alpha\Phi$ and $\mathrm{d}\Phi = 0$, where $\alpha \in \mathbb{R}^*$ (see [26]). An α-Sasakian manifold reduces to a Sasakian manifold (see [23]) when $\alpha = 1$. A normal almost contact metric manifold is called a *β-Kenmotsu manifold* if it satisfies $\mathrm{d}\eta = 0$ and $\mathrm{d}\Phi = 2\beta\eta \wedge \Phi$, where $\beta \in \mathbb{R}^*$ (see [26]). A β-Kenmotsu manifold becomes a Kenmotsu manifold when $\beta = 1$. A normal almost contact metric manifold is called a *cosymplectic manifold* if it satisfies $\mathrm{d}\eta = 0$ and $\mathrm{d}\Phi = 0$.

Putting $Y = \xi$ into (3) and using (2), we have

$$\nabla_X \xi = -\alpha\phi X + \beta(X - \eta(X)\xi) \tag{4}$$

for any vector field X. In this paper, all manifolds are assumed to be connected.

3. Reeb Flow Invariant Ricci Operator on Trans-Sasakian 3-Manifolds

In this section, we give a proof of our main result Theorem 1. First, we introduce the following two important lemmas (see [12]) which are useful for our proof.

Lemma 1. *On a trans-Sasakian 3-manifold of type (α, β) we have*

$$\xi(\alpha) + 2\alpha\beta = 0. \tag{5}$$

Lemma 2. *On a trans-Sasakian 3-manifold of type* (α, β)*, the Ricci operator is given by*

$$\begin{aligned} Q = &\left(\frac{r}{2} + \xi(\beta) - \alpha^2 + \beta^2\right) \mathrm{id} - \left(\frac{r}{2} + \xi(\beta) - 3\alpha^2 + 3\beta^2\right) \eta \otimes \xi \\ &+ \eta \otimes (\phi(\nabla\alpha) - \nabla\beta) + g(\phi(\nabla\alpha) - \nabla\beta, \cdot) \otimes \xi, \end{aligned} \tag{6}$$

where by ∇f *we mean the gradient of a function* f.

We also need the following lemma (see [17])

Lemma 3. *On a trans-Sasakian 3-manifold of type* (α, β)*, the following three conditions are equivalent:*

(1) The Reeb vector field is minimal or harmonic.
(2) The following equation holds: $\phi\nabla\alpha - \nabla\beta + \xi(\beta)\xi = 0$ $(\Leftrightarrow \nabla\alpha + \phi\nabla\beta + 2\alpha\beta\xi = 0)$.
(3) The Reeb vector field is an eigenvector field of the Ricci operator.

Lemma 4. *The Ricci operator on a cosymplectic 3-manifold is invariant along the Reeb flow.*

The above lemma can be seen in [19]

Lemma 5. *The Ricci operator on an* α*-Sasakian 3-manifold is invariant along the Reeb flow.*

Proof. According to Lemma 2 and the definition of an α-Sasakian 3-manifold, the Ricci operator is given by

$$QX = \left(\frac{r}{2} - \alpha^2\right) X - \left(\frac{r}{2} - 3\alpha^2\right) \eta(X)\xi, \tag{7}$$

for any vector field X and certain nonzero constant α. Moreover, according to [16] (Corollary 3.10), we observe that the scalar curvature r is invariant along the Reeb vector field ξ, i.e., $\xi(r) = 0$. In fact, such an equation can be deduced directly by using the formula $\mathrm{div}Q = \frac{1}{2}\nabla r$ and (7). Applying $\xi(r) = 0$, it follows directly from (7) that $\mathcal{L}_\xi Q = 0$. □

Proof of Theorem 1. Let M be a trans-Sasakian 3-manifold and e be a unit vector field orthogonal to ξ. Then, $\{\xi, e, \phi e\}$ forms a local orthonormal basis on the tangent space for each point of M. The Levi–Civita connection ∇ on M can be written as the following (see [12])

$$\begin{aligned} \nabla_\xi \xi &= 0,\ \nabla_\xi e = \lambda\phi e,\ \nabla_\xi \phi e = -\lambda e, \\ \nabla_e \xi &= \beta e - \alpha\phi e,\ \nabla_e e = -\beta\xi + \gamma\phi e,\ \nabla_e \phi e = \alpha\xi - \gamma e, \\ \nabla_{\phi e}\xi &= \alpha e + \beta\phi e,\ \nabla_{\phi e} e = -\alpha\xi - \delta\phi e,\ \nabla_{\phi e} = -\beta\xi + \delta e, \end{aligned} \tag{8}$$

where λ, γ and δ are smooth functions on some open subset of the manifold. We assume that the Ricci operator is invariant along the Reeb flow. From (1) and (4), we have

$$0 = (\mathcal{L}_\xi Q)X = (\nabla_\xi Q)X + \alpha\phi QX - \alpha Q\phi X + \beta\eta(QX)\xi - \beta\eta(X)Q\xi \tag{9}$$

for any vector field X.

By using the local basis $\{\xi, e, \phi e\}$ and Lemma 2, the Ricci operator can be rewritten as the following:

$$\begin{aligned} Q\xi &= \phi\nabla\alpha - \nabla\beta + (2\alpha^2 - 2\beta^2 - \xi(\beta))\xi, \\ Qe &= \left(\frac{r}{2} + \xi(\beta) - \alpha^2 + \beta^2\right) e - (\phi e(\alpha) + e(\beta))\xi, \\ Q\phi e &= \left(\frac{r}{2} + \xi(\beta) - \alpha^2 + \beta^2\right) \phi e + (e(\alpha) - \phi e(\beta))\xi. \end{aligned} \tag{10}$$

Replacing X in (9) by ξ, we obtain

$$\begin{aligned}&\nabla_\xi(\phi\nabla\alpha-\nabla\beta)+\xi(2\alpha^2-2\beta^2-\xi(\beta))\xi+\alpha(-\nabla\alpha+\xi(\alpha)\xi-\phi\nabla\beta)\\&+2\beta(\alpha^2-\beta^2-\xi(\beta))\xi-\beta(\phi\nabla\alpha-\nabla\beta)-\beta(2\alpha^2-2\beta^2-\xi(\beta))\xi=0.\end{aligned}\tag{11}$$

Taking the inner product of the above equation with ξ, e and ϕe, respectively, we obtain

$$\begin{aligned}\xi(\xi(\beta))+2\beta\xi(\beta)+4\alpha^2\beta&=0,\\ \alpha e(\alpha)-\beta\phi e(\alpha)-\beta e(\beta)-\alpha\phi e(\beta)&=0,\\ \beta e(\alpha)+\alpha\phi e(\alpha)+\alpha e(\beta)-\beta\phi e(\beta)&=0,\end{aligned}\tag{12}$$

where we have employed Lemma 1. The addition of the second term of (12) multiplied by α to the third term of (12) multiplied by β gives

$$(\alpha^2+\beta^2)(e(\alpha)-\phi e(\beta))=0.\tag{13}$$

Following (13), we consider the following several cases.

Case i: $\alpha^2+\beta^2=0$, or equivalently, $\alpha=\beta=0$. In this case, the manifold becomes a cosymplectic 3-manifold. The proof for this case is completed because of Lemma 4.

Case ii: $\alpha^2+\beta^2\neq 0$. It follows immediately from (13) that $e(\alpha)-\phi e(\beta)=0$, or equivalently, $g(\nabla\alpha+\phi\nabla\beta,e)=0$. Because e is assumed to be an arbitrary vector field, it follows that $\nabla\alpha+\phi\nabla\beta=\eta(\nabla\alpha+\phi\nabla\beta)\xi$, i.e.,

$$\nabla\alpha+\phi\nabla\beta+2\alpha\beta\xi=0,\tag{14}$$

or equivalently, $\phi\nabla\alpha-\nabla\beta+\xi(\beta)\xi=0$, where we have used Lemma 1. When $\beta=0$, it follows from (14) that α is a nonzero constant. Thus, the proof can be done by applying Lemma 5. In what follows, we consider the last case.

Case iii: $\alpha^2+\beta^2\neq 0$ and $\beta\neq 0$. In this context, (10) becomes

$$\begin{aligned}Q\xi&=2(\alpha^2-\beta^2-\xi(\beta))\xi,\\ Qe&=\left(\frac{r}{2}+\xi(\beta)-\alpha^2+\beta^2\right)e,\\ Q\phi e&=\left(\frac{r}{2}+\xi(\beta)-\alpha^2+\beta^2\right)\phi e.\end{aligned}\tag{15}$$

Replacing X by e in (9) and using (8), (15), we acquire

$$0=(\mathcal{L}_\xi Q)e=\xi\left(\frac{r}{2}+\xi(\beta)-\alpha^2+\beta^2\right)e.$$

With the aid of Lemma 1 and the first term of (12), from the previous relation, we have

$$\xi(r)=0.\tag{16}$$

From (15), we calculate the derivative of the Ricci operator as the following:

$$\begin{aligned}(\nabla_\xi Q)\xi&=0,\\ (\nabla_e Q)e&=e(A)e-\beta A\xi+2\beta(\alpha^2-\beta^2-\xi(\beta))\xi,\\ (\nabla_{\phi e}Q)\phi e&=\phi e(A)\phi e-\beta A\xi+2\beta(\alpha^2-\beta^2-\xi(\beta))\xi,\end{aligned}\tag{17}$$

where we have used the first term of (8) and (12) and, for simplicity, we put

$$A=\frac{r}{2}+\xi(\beta)-\alpha^2+\beta^2.\tag{18}$$

On a Riemannian manifold, we have $\mathrm{div}Q = \frac{1}{2}\nabla r$. In this context, it is equivalent to

$$g((\nabla_{\xi}Q)\xi + (\nabla_{e}Q)e + (\nabla_{\phi e}Q)\phi e, X) = \frac{1}{2}X(r) \tag{19}$$

for any vector field X. Replacing X in (19) by ξ and recalling (16) and the first term of (12), we obtain $2\beta(A - 2\alpha^2 + 2\beta^2 + 2\xi(\beta)) = 0$, or equivalently,

$$\xi(\beta) - \alpha^2 + \beta^2 = -\frac{r}{6}, \tag{20}$$

where we have used the assumption $\beta \neq 0$ and (18). According to (15), it is clear to see that the manifold is Einstein, i.e, $Q = \frac{r}{3}\mathrm{id}$. Because the manifold is of dimension three, then it must be of constant sectional curvature. □

A Riemannian manifold is said to be locally symmetric if $\nabla R = 0$ and this is equivalent to $\nabla Q = 0$ for dimension three. Wang and Liu in [16] proved that a trans-Sasakian 3-manifold is locally symmetric if and only if it is locally isometric to the sphere space $\mathbb{S}^3(c^2)$, the hyperbolic space $\mathbb{H}^3(-c^2)$, the Euclidean space $\mathbb{R}^3$, product space $\mathbb{R} \times \mathbb{S}^2(c^2)$ or $\mathbb{R} \times \mathbb{H}^2(-c^2)$, where c is a nonzero constant. According to [16], on a locally symmetric trans-Sasakian 3-manifold, the Reeb vector field is an eigenvector field of the Ricci operator. Thus, following Lemma 3 and relations (9) and (10), we observe that Ricci parallelism is stronger than the Reeb flow invariant Ricci operator. Hence, our main result in this paper extends [16] (Theorem 3.12).

From Theorem 1, we obtain a new characterization of proper trans-Sasakian 3-manifolds.

Theorem 2. *A compact trans-Sasakian 3-manifold with Reeb flow invariant Ricci operator is homothetic to either a Sasakian manifold or a cosymplectic manifold.*

Proof. As seen in the proof of Theorem 1, a trans-Sasakian 3-manifold with Reeb flow invariant Ricci operator is a α-Sasakian manifold, a cosymplectic manifold or a space of constant sectional curvature. It is well known that an α-Sasakian manifold is homothetic to a Sasakian manifold. Moreover, there do exist compact Sasakian and cosymplectic manifolds. To complete the proof, we need only to prove that *Case iii* in the proof of Theorem 1 cannot occur.

Let M be a trans-Sasakian 3-manifold satisfying *Case iii*. According to (14) and Lemma 5, we know that the Reeb vector field is minimal or harmonic. It has been proved in [17] (Lemma 5.1) that when ξ of a compact trans-Sasakian 3-manifold is minimal or harmonic, then α is a constant. Because the manifold is of constant sectional curvature, then the scalar curvature r is also a constant. Therefore, the differentiation of (20) along ξ gives

$$\xi(\xi(\beta)) + 2\beta\xi(\beta) = 0. \tag{21}$$

Adding the above equation to the first term of (12) implies that $\alpha = 0$ because of $\beta \neq 0$. Using this in (14), we have $\nabla\beta = \xi(\beta)\xi$. The following proof follows directly from [2]. For sake of completeness, we present the detailed proof.

Applying $\nabla\beta = \xi(\beta)\xi$ and (7), we obtain

$$\nabla_X\nabla\beta = X(\xi(\beta))\xi + \xi(\beta)(\beta X - \beta\eta(X)\xi) = 0$$

for any vector field X. Contracting X in the previous relation and using (21), we obtain $\Delta\beta = \xi(\xi(\beta)) + 2\beta\xi(\beta) = 0$. Because the manifold is assumed to be compact, the application of the divergence theorem gives that β is a non-zero constant. Next, we show that this is impossible. In fact, the application of (4) gives that $\mathrm{div}\xi = 2\beta$. Since the manifold is assumed to be compact, it follows that $\beta = 0$, a contradiction. This completes the proof. □

Theorem 2 can also be written as follows.

Theorem 3. *A compact trans-Sasakian 3-manifold with Reeb flow invariant Ricci operator is proper.*

The curvature tensor R of a trans-Sasakian 3-manifold is given by (see [10,27])

$$\begin{aligned} &R(X,Y)Z \\ =&B(g(Y,Z)X - g(X,Z)Y) - Cg(Y,Z)\eta(X)\xi \\ &+ g(Y,Z)(\eta(X)(\phi\nabla\alpha - \nabla\beta) - g(\nabla\beta - \phi\nabla\alpha, X)\xi) \\ &+ Cg(X,Z)\eta(Y)\xi - g(X,Z)(\eta(Y)(\phi\nabla\alpha - \nabla\beta) - g(\nabla\beta - \phi\nabla\alpha, Y)\xi) \\ &- (g(\nabla\beta - \phi\nabla\alpha, Z)\eta(Y) + g(\nabla\beta - \phi\nabla\alpha, Y)\eta(Z))X - C\eta(Y)\eta(Z)X \\ &+ (g(\nabla\beta - \phi\nabla\alpha, Z)\eta(X) + g(\nabla\beta - \phi\nabla\alpha, X)\eta(Z))X + C\eta(X)\eta(Z)Y \end{aligned} \tag{22}$$

for any vector fields X, Y, Z, where, for simplicity, we set

$$B = \frac{r}{2} + 2\xi(\beta) - 2\alpha^2 + 2\beta^2,\ C = \frac{r}{2} + \xi(\beta) - 3\alpha^2 + 3\beta^2. \tag{23}$$

Substituting (14) and (20) into (22), with the aid of (23), we get

$$R(X,Y)Z = \frac{r}{6}(g(Y,Z)X - g(X,Z)Y)$$

for any vector fields X, Y, Z. This implies that, on a trans-Sasakian 3-manifold satisfying *Case iii* in the proof of Theorem 1, we do not know whether $\alpha = 0$ or not. In view of this, we introduce an interesting question:

Problem 1. *Is there a non-proper and non-compact trans-Sasakian 3-manifold of constant sectional curvature?*

Remark 2. *According to De and Sarkar [10] (Theorem 5.1), we observe that a compact trans-Sasakian 3-manifold of constant sectional curvature is either α-Sasakian or β-Kenmotsu.*

Remark 3. *Given a trans-Sasakian 3-manifold, following proof of Theorem 1, we still do not know whether β is a constant or not even when $\alpha = 0$ and the manifold is compact (see [2]).*

Author Contributions: X.L. introduced the problem. Y.Z. investigated the problem. W.W. wrote the paper.

Acknowledgments: This paper was supported by the research foundation of Henan University of Technology. The authors would like to thank the reviewers for their useful comments and suggestions.

Conflicts of Interest: The authors declare no conflict of interest.

References

1. Marrero, J.C. The local structure of trans-Sasakian manifolds. *Ann. Mat. Pura Appl.* **1992**, *162*, 77–86. [CrossRef]
2. Wang, W.; Wang, Y. A Remark on Trans-Sasakian 3-Manifolds. Unpublished work.
3. Aktan, N.; Yildirim, M.; Murathan, C. Almost f-cosymplectic manifolds. *Mediterr. J. Math.* **2014**, *11*, 775–787. [CrossRef]
4. Mangione, V. Harmonic maps and stability on f-Kenmotsu manifolds. *Int. J. Math. Math. Sci.* **2008**, *2008*. [CrossRef]
5. Olszak, Z.; Rosca, R. Normal locally conformal almost cosymplectic manifolds. *Publ. Math. Debrecen* **1991**, *39*, 315–323.
6. Yildiz, A.; De, U.C.; Turan, M. On 3-dimensional f-Kenmotsu manifolds and Ricci solitons. *Ukrainian Math. J.* **2013**, *65*, 684–693. [CrossRef]

7. De, U.C.; De, K. On a class of three-dimensional trans-Sasakian manifolds. *Commun. Korean Math. Soc.* **2012**, *27*, 795–808. [CrossRef]
8. De, U.C.; Mondal, A. On 3-dimensional normal almost contact metric manifolds satisfying certain curvature conditions. *Commun. Korean Math. Soc.* **2009**, *24*, 265–275. [CrossRef]
9. De, U.C.; Mondal, A.K. The structure of some classes of 3-dimensional normal almost contact metric manifolds. *Bull. Malays. Math. Sci. Soc.* **2013**, *36*, 501–509.
10. De, U.C.; Sarkar, A. On three-dimensional trans-Sasakian manifolds. *Extr. Math.* **2008**, *23*, 265–277. [CrossRef]
11. De, U.C.; Yildiz, A.; Yalınız, A.F. Locally ϕ-symmetric normal almost contact metric manifolds of dimension 3. *Apll. Math. Lett.* **2009**, *22*, 723–727. [CrossRef]
12. Deshmukh, S.; Tripathi, M.M. A Note on compact trans-Sasakian manifolds. *Math. Slovaca* **2013**, *63*, 1361–1370. [CrossRef]
13. Deshmukh, S. Trans-Sasakian manifolds homothetic to Sasakian manifolds. *Mediterr. J. Math.* **2016**, *13*, 2951–2958. [CrossRef]
14. Deshmukh, S. Geometry of 3-dimensional trans-Sasakaian manifolds. *An. Stiint. Univ. Al. I Cuza Iasi Mat.* **2016**, *63*, 183–192. [CrossRef]
15. Deshmukh, S.; Al-Solamy, F. A Note on compact trans-Sasakian manifolds. *Mediterr. J. Math.* **2016**, *13*, 2099–2104. [CrossRef]
16. Wang, W.; Liu, X. Ricci tensors on trans-Sasakian 3-manifolds. *Filomat* **2018**, in press.
17. Wang, Y. Minimal and harmonic Reeb vector fields on trans-Sasakian 3-manifolds. *J. Korean Math. Soc.* **2018**, *55*, 1321–1336.
18. Cho, J.T. Contact 3-manifolds with the Reeb-flow symmetry. *Tohoku Math. J.* **2014**, *66*, 491–500. [CrossRef]
19. Cho, J.T. Reeb flow symmetry on almost cosymplectic three-manifolds. *Bull. Korean Math. Soc.* **2016**, *53*, 1249–1257. [CrossRef]
20. Cho, J.T.; Kimura, M. Reeb flow symmetry on almost contact three-manifolds. *Differ. Geom. Appl.* **2014**, *35*, 266–273. [CrossRef]
21. Cho, J.T.; Chun, S.H. Reeb flow invariant unit tangent sphere bundles. *Honam Math. J.* **2014**, *36*, 805–812. [CrossRef]
22. Suh, Y.J. Real hypersurfaces in complex two-plane Grassmannians with ξ-invariant Ricci tensor. *J. Geom. Phys.* **2011**, *61*, 808–814. [CrossRef]
23. Blair, D.E. *Riemannian Geometry of Contact and Symplectic Manifolds*; Springer: Berlin, Gernamy, 2010.
24. Olszak, Z. Normal almost contact metric manifolds of dimension three. *Ann. Polon. Math.* **1986**, *47*, 41–50. [CrossRef]
25. Chinea, D.; Gonzalez, C. A classification of almost contact metric manifolds. *Ann. Mat. Pura Appl.* **1990**, *156*, 15–36. [CrossRef]
26. Janssens, D.; Vanhecke, L. Almost contact structures and curvature tensors. *Kodai Math. J.* **1981**, *4*, 1–27. [CrossRef]
27. De, U.C.; Tripathi, M.M. Ricci tensor in 3-dimensional trans-Sasakian manifolds. *Kyungpook Math. J.* **2003**, *43*, 247–255.

MDPI
St. Alban-Anlage 66
4052 Basel
Switzerland
Tel. +41 61 683 77 34
Fax +41 61 302 89 18
www.mdpi.com

Mathematics Editorial Office
E-mail: mathematics@mdpi.com
www.mdpi.com/journal/mathematics

www.ingramcontent.com/pod-product-compliance
Lightning Source LLC
Chambersburg PA
CBHW041217220326
41597CB00033BA/5999
9783039218004